T0342519

Human Programming

Human Programming

*Brainwashing, Automatons, and
American Unfreedom*

S<small>COTT</small> S<small>ELISKER</small>

UNIVERSITY OF MINNESOTA PRESS
MINNEAPOLIS · LONDON

Chapter 2 was previously published as "Simply by Reacting? The Sociology of Race and *Invisible Man*'s Automata," *American Literature* 83, no. 3 (2011): 571–96. Reprinted by permission of Duke University Press.

Typescript selections from Ralph Ellison's unpublished "Invisible Man" are reproduced in chapter 2. Copyright 1951, 2015 by Ralph Ellison. Courtesy of the Ralph Ellison Papers at the Manuscript Division of the Library of Congress; reprinted by permission of The Wylie Agency LLC.

Copyright 2016 by the Regents of the University of Minnesota

All rights reserved. No part of this publication may be reproduced, stored in a retrieval system, or transmitted, in any form or by any means, electronic, mechanical, photocopying, recording, or otherwise, without the prior written permission of the publisher.

Published by the University of Minnesota Press
111 Third Avenue South, Suite 290
Minneapolis, MN 55401-2520
http://www.upress.umn.edu

Printed in the United States of America on acid-free paper

The University of Minnesota is an equal-opportunity educator and employer.

23 22 21 20 19 18 17 16   10 9 8 7 6 5 4 3 2 1

Library of Congress Cataloging-in-Publication Data
Names: Selisker, Scott, author.
Title: Human programming : brainwashing, automatons, and American unfreedom / Scott Selisker.
Description: Minneapolis : University of Minnesota Press, 2016. | Includes bibliographical references and index.
Identifiers: LCCN 2015036887 | ISBN 978-0-8166-9987-2 (hc) | ISBN 978-0-8166-9988-9 (pb)
Subjects: LCSH: Brainwashing—United States. | Psychological warfare—United States. | Robots—United States.
Classification: LCC BF633 .S45 2016 | DDC 153.8/530973—dc23
LC record available at http://lccn.loc.gov/2015036887

Contents

Acknowledgments

For guidance and rich conversations in the earlier phases of the project, I thank Jennifer Wicke, Eric Lott, David Golumbia, David Sigler, Erich Nunn, Walt Hunter, Chris Forster, Wes King, Nathan Ragain, Shaun Cullen, Gabriel Hankins, and Greg Colomb. For generous pointers, references, or questions that had a significant impact on the final text, I thank Priscilla Wald, Swati Rana, Enda Duffy, Vaughn Rasberry, Deak Nabers, Jed Esty, Jennifer S. Rhee, Karim Mattar, and Doug Armato. And most especially, for their incisive responses to earlier drafts of the present manuscript, I express my gratitude to Faith Harden, Nathan K. Hensley, Tim Wientzen, Natalia Cecire, Andrew Griffin, Josh Epstein, Katie Fitzpatrick, Anne Garland Mahler, Paul Hurh, and Laura Goldblatt. For their generous engagements with the manuscript as a whole, I thank Mark Goble and an anonymous reader for the University of Minnesota Press.

On an institutional level, this work was supported by a New Faculty Fellowship from the American Council of Learned Societies and the Andrew W. Mellon Foundation in 2012–13 and 2013–14, and I'm grateful to Alan Liu, Rita Raley, and the English department at the University of California, Santa Barbara for hosting me in that fellowship. I'm also grateful to Larry Evers, Lee Medovoi, and my other colleagues in the Department of English at the University of Arizona for their generous support during my transition to that institution. I'm also indebted to the

University of Virginia for fellowship support from 2004 to 2010, and I'm thankful for suggestions from conference audiences at the Modern Language Association, the American Studies Association, the Society for Novel Studies, and the Society for Literature, Science, and the Arts and from workshop coparticipants at the Dartmouth Futures of American Studies Institute and the New England Americanist Collective. Undergraduate students in my I/Robot course at UCSB and in my Automaton course at the University of Arizona also helped me to hone the ideas herein.

My deepest debt of gratitude, in writing this book as in so many other things, is to Faith Harden.

Enemies of Freedom

In November 2001, an American citizen named John Walker Lindh was taken prisoner by the United States during combat with the Taliban. This "American Taliban" was featured on magazine covers, in headlines, and in ongoing news coverage across the country well into 2002. The coverage revolved around versions of a single question: how could a well-off young American become an enemy combatant, fighting for a group allied with the terrorists behind 9/11? Just over a month before Lindh's capture, George W. Bush described the 9/11 hijackers as "enemies of freedom," and he issued the well-known ultimatum "either you are with us, or you are with the terrorists" as a strategy for consolidating national sentiment in the face of the 2001 terrorist attacks.[1] In the year of the American flag bumper sticker, Lindh had made a seemingly impossible choice: he was with the terrorists. In news stories that purported to help readers understand that decision, shocked friends and family members frequently wondered aloud how the Taliban "brainwashed" him.[2] The major U.S. newspapers ran series of articles and commentaries whose coverage of Lindh reached out in several directions for historical precedents. Interviewees in the *Los Angeles Times* included experts such as Philip Zimbardo (then head of the American Psychological Association and best known for the 1971 Stanford Prison Experiment) and the families of some of the first Americans to be called "brainwashed," the twenty-one POWs who refused to repatriate at the end of the Korean War. One expert in favor of a brainwashing defense for Lindh, the counselor Steven Hassan, MD, speculated that, "with proper counseling, Lindh would be horrified at what brainwashing has made him do. . . . Indeed, it appears that mind control techniques . . . have been used by Osama bin Laden and his cronies to recruit, train, and exploit talented people to do their bidding."[3] A close friend of Lindh's family stated, "We feel it was almost like a Patty Hearst thing," referring to the newspaper heiress who first attempted a brainwashing legal defense in 1976.[4] This metaphor of brainwashing, at once familiar and strange, stuck; soon after, the idea of the brainwashed also appeared in families' descriptions, and the legal defense strategies, of other home-grown terrorists: the American enemy combatant José Padilla; Washington, D.C., sniper Lee Boyd Malvo; the British "shoe bomber" Richard Reid; the convicted 9/11 conspirator Zacarias Moussaoui, and others.[5] In spring 2015, a front-page story in the *New York Times* mentioned brainwashing in a story about a young man from Minnesota, Abdi Nur, who had joined the fundamentalist state of ISIS.[6] The hit cable television series *Homeland* (2011–) featured an American convert to Islam who closely resembled Lindh, and numerous other books in popular and literary culture have probed the psychological dimensions of religious conversion, torture, and coercion since 9/11 through the brainwashing trope. Those young men's families imagined the choice of joining the terrorists as unthinkable, even for the young men themselves. The narrative of brainwashing had significant explanatory power, as it attributed Lindh's unthinkable choice to the shadowy operations of indoctrination and manipulation on the part of the fundamentalist leaders and to what I will refer to as *automatism* in the ordinary human subject. Moreover, this image of brainwashing in the popular press allows readers to see themselves as comparatively free thinkers in a democratic society, vis-à-vis the apparent unfreedom of democracy's enemies.

Lindh's case reveals that our ways of talking about freedom and unfreedom in the United States are suffused with the intertwined histories of psychology and philosophy, of political theory, communications technology, literature, and film. *Human Programming* explores this story, and others like it, by charting the development of a cinematic, literary, and

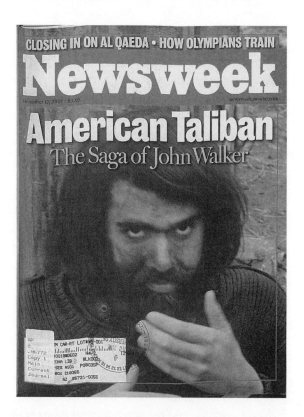

Above: John Walker Lindh, in a cover story for *Newsweek* on December 17, 2001. This "American Taliban" member was the subject of extensive news coverage, much of which included speculation on thought reform and "brainwashing." *Below*: Damian Lewis as Nicholas Brody, shown in al-Qaeda captivity, in the Showtime series *Homeland* (2011–). Some elements of Brody's story and appearance resonate with Lindh's, while others echo fictions of mind control such as *The Manchurian Candidate* (Frankenheimer, 1962).

rhetorical figure I call, borrowing from Erich Fromm, the *human autom-*
aton. The human automaton is the image of an individual deprived of
his or her free will, that is, programmed or "brainwashed" into acting or
speaking without conscious control, and it's an image that recurs not
just in science fiction but in news coverage, political speeches, and social
science articles and books. The human automaton recurs in televisual
images of the brainwashed terrorist, accounts of the dissenting Korean
War POWs to whom the term *brainwashing* was first applied, propaganda
films of the totalitarian enemy in World War II, partisan news commentary, and descriptions of the psychology of cult members or extremists.
It's the range and political seriousness of these representations that first
convinced me not to dismiss the rhetorical popularity of "brainwashing"
as merely the afterlife of debunked pseudoscience. Rather, in an argument I pursue in these pages, the human automaton has played a key
role in American descriptions of geopolitical difference and diversity
since World War II.

This figure of the human automaton has given Americans over the
past seventy-five years a stable way to talk about nebulous differences:
ideology, religion, and the psychological states of individuals who make
what seem to be impossible choices. On partisan news networks as much
as in propaganda and science fiction, it seems easiest to think of others
who deeply disagree with our own viewpoints as victims of false consciousness, as unthinking masses, or as easily swayed "sheeple." There
are, of course, major differences, to which we will attend, between the
psychological state attributed to a John Walker Lindh or a Patty Hearst
and the kinds of social and political realities of Hitler's Germany or
Mao's China, such as repressive censorship and secret police. But even
George Kennan, chief policy architect of the U.S. Cold War, has noted
the power that the figurative face of mental unfreedom has in the American imagination. "When I try to picture totalitarianism to myself as a
general phenomenon," Kennan said to the American Academy of Arts
and Sciences in 1953, "what comes to mind most prominently is neither
the Soviet picture nor the Nazi picture as I have known them in the flesh,
but rather the fictional and symbolic images created by such people as
Orwell or Kafka or Koestler or the early Soviet satirists."[7] Likewise, the

image of Lindh as brainwashed came to his friends and family from popular texts and narratives. For Kennan, the "purest expression of the phenomenon" of totalitarianism is "rendered not in its physical reality but in its power as a dream, or nightmare." It is in the "state of mind it creates in its victims," Kennan says, that "totalitarianism reveals most deeply its meaning and nature."[8] *Human Programming* sets out to understand how and why totalitarians and other "enemies of freedom" have been imagined so readily through the human automaton's unfree "state of mind."

Our ways of imagining that state of mind have been shaped by developments in science and technology and by conventions in fiction, film, and other popular cultural texts. In the typical fictional narrative about a human automaton, a new kind of technology or scientific knowledge puts a would-be dictator on the cusp of controlling individuals in a politically catastrophic way, leaving free characters to fight against the threat of unfreedom, sometimes even within their own minds. On film, the mechanisms of unfreedom are often exaggerated for visual effect, as they are in John Frankenheimer's *The Manchurian Candidate* (1962), where we watch the brainwashed sleeper agent Raymond Shaw transition from his ordinary state to his programmed state and back again with stunning regularity. We also see, by contrast, Frank Sinatra's character, Captain Marco, consistently struggling against the programming in his own brain by waking up from his nightmares and remembering his programming. Through this struggle with unfreedom, he dramatizes his free will. Freedom from slavery, from prison, or from tyranny is easy enough to represent on film or in literature, but the freedom of a free mind is a more difficult proposition. Indeed, it's only by contrast with Shaw's unfreedom that Marco's freedom becomes visible. Similar science fictional images of mental unfreedom proliferated widely over the course of the twentieth century, with varying degrees of scientific justification. Many of these images closely followed or extrapolated from the scientific and technological developments that threatened to uncover and then manipulate the deepest and most human characteristics of individuals and groups: behaviorist psychology and Taylorist industrial management beginning in the 1910s, radio and cinematic propaganda that would be theorized throughout the 1930s and 1940s, and the developments of cybernetics and computer science that redefined the differences between humans and machines beginning in the 1940s. Across these paradigms of technoscientific development, the prospect of total mental unfreedom loomed just over successive horizons of scientific possibility, and it is this specter of unfreedom that I suggest accounts for automatism's rhetorical power in the postwar United States.

In arguing that the human automaton has been a major and pervasive figure in U.S. cultural and political life since World War II, *Human Programming* traces a network of exchange between literature, film, science, and political culture that is remarkably omnidirectional. Rather than either reflecting or resisting the scientific and political texts that surrounded them, novels and films were part of an interdisciplinary conversation through which the automaton image emerged. If we take a part of that network around the term *brainwashing* that resurfaced in the War on Terror, we find one such conversation. The term had been coined by Edward Hunter, a propagandist and sometime CIA operative, during the Korean War. Although he claimed to have borrowed the term from the Chinese communists, Hunter's understanding of behavioral conditioning and politics was influenced instead, as he admitted in his later writings, by George Orwell's and Ayn Rand's ideas and fiction.[9] Despite its quasi-literary origins, the problem of brainwashing was studied in the 1950s and 1960s by scientists, including the Columbia psychiatrist Joost Meerloo and the Harvard psychiatrist and social psychologist Robert Jay Lifton, whose work would inform sensationalist media coverage of Patty Hearst and legal strategies related to cults in the 1970s. Looking backward, Orwell's ideas about totalitarian language and behavior closely resembled those of the behaviorist psychologists, including John B. Watson, from whose famous filmed "Little Albert" experiment Orwell borrowed a key scene in *1984*. Moving forward again, *1984* and other antitotalitarian texts would provide a template for progressive writers in the 1950s, 1960s, and 1970s—such as Ralph Ellison, Ken Kesey, and Betty Friedan—to imagine the institutions of the U.S. establishment as totalitarian, their inhabitants as humans reduced to automatons. That association between automatism and institutions floated into the antipsychiatry movement, literary and cultural theory, and discourses about computation and technological posthumanism. And, again, early in the War on Terror, "brainwashing" was revived in 2002 in both interviews with insurgents' families and in courtrooms, to the extent that the *New York Times Magazine* included "the brainwashing defense" in its retrospective of the year in ideas.[10] Following that cluster of brainwashing cases, popular film and television productions, including Jonathan Demme's remake of *The Manchurian Candidate* (2004), *Battlestar Galactica* (2004–9), and *Homeland,* have rehearsed the brainwashed-terrorist scenario in numerous permutations. *Battlestar Galactica* and *Homeland,* moreover, explore the relationships between our culture's images of political extremism and those of religious fundamentalism (a question this book's two final chapters examine as it approaches the present moment). Through this admittedly dizzying variety of situations, the human automaton has provided a visual and narrative grammar through which Americans have discussed pressing questions about mental unfreedom and its geopolitical consequences.

The task of *Human Programming* is to trace some of the applications of that grammar and suggest how and why it works. By focusing on the human automaton as a figure within the social sciences, political discourse, fiction and film, this book seeks to open several new avenues for understanding U.S. culture since 1945. First, it offers a new literary and cultural context for understanding the human automaton figure. Long histories of automatons and robots—such as Minsoo Kang's excellent *Sublime Dreams of Living Machines*[11]—often end with the human robot as a sign of modern factory alienation, as in Fritz Lang's *Metropolis* (1927) or Karel Čapek's *R.U.R.* (1923),[12] but the period of its resurgence in U.S. political culture is just getting started at that moment, when 1920s robots and zombies will be applied to new political and representational crises. (Speaking of robots and zombies, they do appear from time to time in these pages, but it's the human automaton, the person reduced to a machine, that takes center stage here.) Unlike those who have argued that automatons have a single meaning across historical contexts, and others who have argued that automatons in individual texts definitively represent one particular latent anxiety (that Jack Finney's "body snatchers" *are* communists or white suburbia or even migrant labor), I focus instead on how this figure has circulated in postwar culture.[13] The human automaton has been adapted for use on the political right and left, so its form doesn't have an underlying conservative or progressive meaning in the way that symptomatic readers might argue. The automaton and human automaton do, as Kang, Lisa Zunshine, Bill Brown, and Despina Kakoudaki have variously argued, derive their compelling aesthetic effects from moments of ontological uncertainty wherein our categories for personhood and objecthood become confused.[14] But it is the particular pattern of circulation and the uses of the image—again, on both the political left and right—that give it a political importance that's worth attending to in the United States. (As for other cultural contexts, one need only look to the roles of automatons and robots in Japanese culture—from the centuries-old tradition of bunraku puppetry to Masahiro Mori's contention that the robot has a "Buddha nature"—to be certain that automatism has culturally specific meanings and circulations elsewhere that exceed our framework here.)[15] Rather than risk overstating the political impact of literary and cinematic texts, I try to show literary and cinematic texts' political impact by looking closely at their saturation within U.S. culture. In a wide array of domains, these texts have shaped the American vocabulary for talking about politics, will, freedom, and unfreedom.

Second, the technological side of this cultural history has the potential to resituate academic conversations about technological posthumanism, or the difference between human and posthuman being, within a new archive. Since Donna Haraway's "A Cyborg Manifesto,"[16] both academic and popular posthumanists and transhumanists have focused on the expansive possibilities of posthuman futures by comparison with the liberal–humanist present or past: changing the body or mind through technological modifications or interfaces is a way of seeing how the boundaries of human beings are permeable in exciting and sometimes unsettling ways. Academic posthumanists often contrast their ideas of a constructed, networked, posthuman subject against the undifferentiated naïveté of "liberal humanism."[17] But when we consider, as Mark Greif has suggested, that the 1950s liberal humanism to which posthumanists refer was already a response to the threat of a totalitarian automatism, we can reimagine the terms of a humanist–posthumanist divide in a dramatically different light. Rather than a human past and a posthuman future in either physical or philosophical terms, *Human Programming* finds after 1945 a continually developing dialectic, and various geopolitical mappings, between liberal–humanist selves and posthuman—or, more pointedly, subhuman—automatic others.[18] (Looking at the post–Cold War moment in China and the developing world, Shu-mei Shih has also speculated that the dynamic between humanism and posthumanism is a geographic rather than a temporal one.)[19] That mapping of self and other helps us to understand how and why liberal humanist values have persisted and where the posthumanist embrace of a constructed subjectivity might still be a productive gesture. It also situates the history of humanist and posthumanist thought as more central to postwar, mainstream U.S. political culture than posthumanist scholars had previously imagined.

The biggest question at stake in *Human Programming*, though, is how this figure of the human automaton has shaped American conversations about the self and the other, the free and the unfree, democracy and its enemies, since World War II. These enemies, described in the chapters to come, will include totalitarianism and communism as well as the political and religious extremisms of cults and fundamentalist terrorists. In what remains of this introduction, I situate the human automaton in several ways: first, within a thumbnail scientific, literary, and cinematic prehistory; second, within a history of interpretations of its aesthetic effects; and third, within the lineage of American representations of racial and ideological others.

The Art and Science of Automatism

A major debate in psychology at the end of the nineteenth century centered on the question of humans' automatism, or automatic nature. In the "conscious-automaton" theory, which had come into vogue through the work of Thomas Huxley and others, the conscious mind is in fact a passive observer of essentially mechanical bodily processes such that subjective free will is an illusion. As William James describes this theory, the conscious mind is "a steam-whistle which accompanies the work of a locomotive engine."[20] (In the twenty-first century, cognitive philosophers, including D. M. Wegner, have come to strikingly similar conclusions.)[21] In the essay "Are We Automata?," and later in *The Principles of Psychology,* James developed an apparently solid middle ground in the debate. He admits that "ninety-nine hundredths, or, possibly, nine hundred and ninety-nine thousandths of our activity is purely automatic and habitual";[22] in a "habitual action, mere sensation" guides our movements so that "the upper regions of the brain and mind are set comparatively free."[23] Where free will exists for James is as a kind of "selective agency" of consciousness, wherein the exercise of will is described as holding an absent thing present in a person's mind.[24] Even as carefully circumscribed as this, however, the will itself, as Immanuel Kant observed in his first *Critique,* is something that cannot be observed, much less measured.[25]

In the early twentieth century, James's paradigm for doing psychology was overturned by a program that did away with understanding the will altogether to focus on observing and then modifying human habits. Beginning with the 1914 talk "Psychology as the Behaviorist Views It," John B. Watson's advocacy, experiments, and textbooks solidified the experimental program that established psychology as a discipline independent of philosophy. That independence takes a striking form in Watson's psychology: he insisted on the absolute nonexistence of the mind as an object of inquiry. Behavior, he claimed, was the only thing that could be objectively observed in his newly founded discipline. James foresaw the philosophical dimensions of this shift: "when we look at living creatures from a certain point of view," he writes, "one of the first things that strike us is that they are bundles of habits."[26] Watson simply embraces this point of view—an absolute distance from the subject's interiority— and decides not to stray from it. By scientifically observing modulations in behavior, the patterns of learning behavior, and, most important, the *malleability* of behavior, Watson produces a subject that is essentially the engineer's black box: we neither know nor care what is inside, so long as we can make use of its inputs and outputs.[27] The early method of the behaviorists was to focus on conditioning the involuntary reactions—

the dilation of pupils, salivation, and so forth—that remove the subject's conscious intentions from the observation entirely. Speaking of salivation, Watson worked at the same time as the better-known Ivan Pavlov, who is better known in part because of later U.S. anti-Soviet propaganda to which we turn in chapter I. Watson's equivalent of Pavlov's slavering dogs, in terms of notoriety, was the "Little Albert" experiment, in which Watson attempted to condition fear in an infant to coincide with seeing different white animals, including rats. His goal, in this experiment that responded to Freud's "Little Hans" case study, was to show that any behavior, even supposedly deep characteristics such as fear reactions, could be programmed.[28] Overall, behaviorist psychology, by bracketing the will and consciousness, makes the human objects of psychology's inquiry functionally, if not essentially, automatons.

Watson's advocacy for the social utility of his insights, moreover, made behaviorist psychology an early player in twentieth-century technocracy, the rising influence of experts in political and industrial culture that would come to define mid-century American culture.[29] For his part, Watson worked early on to implement a personality test (a precursor to the Armed Services Vocational Aptitude Battery still in use) to determine officer assignments, and his advocacy led to the establishment of the Division of Psychological Warfare during World War II.[30] Watson's behaviorist psychology emerged just three years after Frederick Winslow Taylor's system of scientific management, which sought to optimize workers' bodily movements for every work-related task. Both methods were explicitly used in the service of more efficient management of human labor, designing human workers in the image of machines. This pragmatic, black-box approach to human subjects took shape alongside behaviorist psychology as the malleability and micromanagement of movement itself. The ubiquity of this approach in the early twentieth century was in part responsible for the rise of *mass* automatism as a political question in the twentieth century. The roughly simultaneous advent of cinema and radio, and of programmatic propaganda during World War I, made the possible deception of the masses a political problem that was often discussed in tandem with automatism, as in Orwell's *1984*.

These modern developments in scientific automatism took shape alongside a much longer aesthetic tradition of automatons and automatism. Automatism in theater and dance had been a particularly popular theme from the mid-nineteenth through the early twentieth centuries, in works such as Léo Delibes's and Arthur Saint-Léon's 1870 ballet *Coppélia,* based on E. T. A. Hoffmann stories; Igor Stravinsky's 1911 ballet *Petrushka*; and Oscar Schlemmer's *Triadic Ballet* (1922).[31] Stravinsky's Petrushka doll—its stuttering movements played masterfully by Vaslav Nijinsky—hearkened back to the Pierrot doll, a figure from the commedia dell'arte that also saw a revival on the vaudeville stage of the Ziegfeld Follies.[32] These acts in theater and dance coincided with a turn-of-the-century resurgence of interest in mechanical automatons, which had been popular curiosities, often crafted by watchmakers, in Europe and America since the seventeenth century.[33] Mechanical automatons had long had a purchase in the question of what makes a human, appearing in literature and philosophy from the time of Descartes's mechanistic dualism, in Julien Offray de La Mettrie's "Man a Machine" (1748), and in the eerie mechanical doubles and monstrosities in a long arc of narrative works such as Edmund Spenser's *The Fairie Queene* (1596), E. T. A. Hoffmann's "The Sandman" (1816), and Herman Melville's "The Bell-Tower" (1855). In the royal courts of Europe, the curiosities of Jacques de Vaucanson's "Defecating Duck" and the writing automatons of Henri Maillardet and Pierre Jaquet-Droz had been considered amazing entertainments since the eighteenth century. Through much of the nineteenth century, Wolfgang von Kempelen's chess-playing Turk "automaton" toured Europe and the United States (with a dwarf hidden inside the machine), to appear in the writings of Edgar Allan Poe and, later, Walter Benjamin.

The themes of those artistic representations of automatons and human automatons varied according to medium and purpose. In dance, for instance, automaton and marionette figures most often provided simple conceits for exploring the mechanics of movement and the dynamic nature of choreography within a narrative. Whereas Hoffmann famously achieved uncanny effects in fiction with the automaton doll Olympia,

eighteenth- and nineteenth-century prose descriptions of human automatism more frequently suggested satire, from girls' supposedly empty
imitations of fashion to church officials and French royalty who are—
without anyone noticing!—replaced by automatons.[34] Mary Shelley may
have borrowed the name "Frankenstein" from a French revolutionary
satire about an automaton maker, and François Villiers de l'Isle-Adam
wrote a novel in which Thomas Edison invented a mechanical woman in
1887.[35] The late nineteenth-century fad for mesmerism fell easily within
this satirical framework, and Henry James's *The Bostonians* (1886) includes a scene in which a young feminist's inspiring and idealistic
speeches are the result of a mesmeric laying-on of hands. Automaton
themes and imagery suffuse the poetry of T. S. Eliot, from his early
poems about bourgeois people as empty-headed puppets to the empty
motion that haunts his mature work, such as the "automatic hand" that
appears in the rape scene of "The Waste Land."[36] Eliot, D. H. Lawrence,
and many other modernists depicted modern forms of emptiness through
various resemblances to puppets and machines, and they shared a desire to differentiate, with certainty, between the emptiness of repetitive
gesture and the vitality that underlies genuine agency. While habit and
its manipulability became a tool for psychology and industry, artistic
ideas about habit would set up an opposition between, on one hand,
habit, automatism, and deadness and, on the other, renewal, spontaneity, and humanity. Writers and cultural critics, including Rebecca West,
Theodor Adorno, Clement Greenberg, and George Orwell, would, by the
middle of the century, forge various connections between the modernist
revolt against twentieth-century modernity and the cause of democracy
against totalitarianism.

The aesthetic effects of automatic movement were dramatically heightened within the realm of cinema. Whereas Mark Seltzer has argued that
American fiction at the turn of the twentieth century often rehearsed the
"melodrama of uncertain agency" in its plots about social institutions, cinema's depictions of automatic-looking human motion produced greatly
condensed versions of that melodrama.[37] The "cinema of attractions"
of the first decade of the twentieth century, for instance, incorporated

approximately a dozen short features centered around apparently automatic people, including butlers, policemen, husbands, and more, capitalizing on the peculiar appearance of motion in the still-novel moving picture.[38] This fascination with motion and the automatic cuts across many divisions within early cinema, from the primitive cinema of attractions to later Hollywood films, from intellectual avant-garde films to slapstick and horror.[39] In the 1910s, 1920s, and 1930s, automatic movement drove the visual suspense of Hollywood "B" features—including numerous waxworks sequences, the Frankenstein and Dracula enterprises, *The Mask of Fu Manchu* (Charles Brabin, 1932), *White Zombie* (Victor Hugo Halperin, 1932), and other early zombie films—as well as expressionist masterpieces such as *The Cabinet of Doctor Caligari* (Robert Wiene, 1920), Fritz Lang's *Metropolis* and *Doktor Mabuse* films in the early 1930s, and avant-garde experiments, including aspects of *Un chien andalou* (Luis Buñuel, 1929) and *Ballet mécanique* (Fernand Léger and Dudley Murphy, 1924).[40] The wide variety of uses of automatism in cinema would, as I describe toward the end of this introduction, provide an inventory of aesthetic effects that made possible the cinematic representability of totalitarianism. (I should note that although the word *automatism* in film studies often refers to the automatic operation of the camera, I use it here as a term of art in psychology to describe the phenomena of automatic and apparently programmable motion and speech.) Even after dozens of campy science fiction films have used automatic motion for the cheap special effect that it is, the aesthetic appeal of automatic-seeming human motion has yet to be worn out. The ontological uncertainty that human automatism brings about can still function as a narrative motor, as it has in recent films such as Shane Carruth's *Upstream Color* (2013) and Alex Garland's *Ex Machina* (2015), as well as in television series including Joss Whedon's *Dollhouse* (2009–10) and Melissa Rosenberg's *Jessica Jones* (2015).

While they appear prominently and persistently in literary and cinematic texts themselves, human automatons also made a key appearance in literary and cultural theory through Michel Foucault's *Discipline and Punish*.[41] In this widely influential book, the central image of the institutionalized "docile body" emerges from a scene of fascination with automatons: Foucault recalls how Frederick II of Prussia was "obsessed" with the "celebrated automatons" of the eighteenth century, not only as "a way of illustrating an organism" but also as "political puppets, small-scale models of power."[42] They function for this section of *Discipline and Punish*, too, as models of power over a subject that, as Foucault describes it, "has become something that can be made; out of a formless clay, an inapt body, the machine required can be constructed; posture is gradually corrected; a calculated constraint runs slowly through each part of the body, mastering it, making it pliable, ready at all times, turning silently into the automatism of habit."[43] In this passage, Foucault finds in Western culture since the late eighteenth century the kinds of instrumental reason that thinkers like George Orwell and Erich Fromm, as I show in chapter 1, had associated with totalitarianism. Although Foucault, Giorgio Agamben, and others have further developed the concept of "biopolitics," I have found myself hesitant to overestimate the critical distance it can provide to the subject at hand.[44] Where, for instance, the critic David Seed writes that the 1950s idea of "brainwashing" is an apt "illustration" of Foucault's ideas about discipline (and Louis Althusser's, on ideology) from the 1970s, I'm inclined to wonder if it's not at least in part the other way around. I find in chapter 2 many instances where the anti-totalitarian automatism rhetoric of the 1950s became the source of powerful automaton images within 1960s progressivism and the antipsychiatry movement. Foucault, a major figure of the latter groups, complimented Deleuze and Guattari's *Anti-Oedipus*[45] by suggesting it could be titled *Introduction to the Non-fascist Life,* one sign among others that his work is, naturally, a product of its postwar intellectual milieu.[46] Accordingly, although I borrow from Foucault's insights on occasion, my methodology focuses on examining the confluence of aesthetics and politics in the automaton's effects, following Jacques Rancière, and on tracing its exchanges between disparate domains, following Bruno Latour. But the presence of Foucault's automatons in the tradition of critical theory can leave critics and historians of culture with several self-reflexive questions that *Human Programming* addresses within the cultural artifacts it takes up: when and why do we assume that institutions hold the power to eliminate individuals' abilities to make free choices? How have dialectics of self and other infused even the most sophisticated critical accounts of false consciousness, which have persisted from Plato's Cave through the tradition of Western Marxism? How often do versions of Eliot's bourgeois puppets populate critical accounts of consumption and citizenship, and what roles do they play in the rhetoric of contemporary critical theory? To gain some purchase on these questions, we must first describe the aesthetic dimensions of automatism.

The Aesthetics of Automatism

We require, then, a theory of automatism that makes sense of its versatility in representing others. In developing this theory, I want to develop a sense of automatism's *aesthetics* in the special sense that the political philosopher Jacques Rancière uses the word. Taking the Greek root word *aisthêsis* in its broader sense of "perception," Rancière notes that the way we constitute or count political communities is an aesthetic matter, that is, one that depends on the categories around which we train our perception.[47] For Rancière, the most important dimension of democratic politics is not that of individuals quarrelling around a table but rather the divisions through which some groups' voices find a spot at the table and others' do not. Thus politics is a "quarrel over the constitution of the *aisthêsis*, over this partition of the perceptible through which bodies find themselves in community."[48] I find in the early twentieth-century analyses of automatism, by Henri Bergson and Sigmund Freud, a way of seeing just that kind of aesthetics of human community and its boundaries. The irrational effects of fright and laughter, in their analyses, can reveal to us the kinds of individuals whom we perceive as inhabiting the boundary between humanity and objecthood.

On the screen and on the page, human automatons can be used for humorous moments, or they can be used to create the horror effects that Sigmund Freud, and later Masahiro Mori, would call "uncanny" feelings of unease and fright. The comedic and uncanny effects of movement on the screen were wildly popular in early cinema, and the early twentieth-century theories of these effects suggest, provocatively, that mere movement might have political and ethical significance. One of the richest texts on comedic effects, the phenomenologist Henri Bergson's essay "Laughter," ties the effects of comedic motion to notions of community and vitality.[49] In an evocative image, Bergson postulates that visual comedy is fundamentally tied to automatism, as "something mechanical encrusted on the living."[50] In this general theory of laughter, which encompasses verbal humor, situational comedy, and physical slapstick, the butt of the joke is shown to be behaving in a mechanical and, consequently, sub-human fashion. The mechanicity is in the eye of the beholder, who perceives a changed situation to which the subject has not yet adapted: it might be classed norms of behavior, as in the stock situation of the lower-class ruffians who try, but fail, to adapt to the etiquette of the fancy restaurant, or it might be a purely bodily motion, such as Wile E. Coyote continuing to vibrate after using a jackhammer. Charlie Chaplin's *Modern Times* (1936) provides an instructively literal example of Bergsonian mechanism. The Tramp works at a conveyor belt in which he uses wrenches to tighten all the hexagonal bolts that pass by on the belt, at a near-superhuman speed. He falls into the machine itself and seems to be possessed all the more with a bolt-tightening mania, a mental automatism. When he leaves the factory, his arms continue to twitch with his tightening motions, and he immediately comes across a woman whose dress features hexagonal buttons across the bust. Hilarity ensues as the woman flees the Tramp, who persists in his attempt to be a human bolt-tightening machine. The Tramp's behavior here is precisely the sort of "mechanical encrusted on the living" that Bergson has described: his behavior as a bolt-tightener continues after he has walked outside the factory, and not only his motion but his perception—in imagining all hexagons as bolts to be tightened—remains in "factory" mode. Indeed, part of the humor comes from the uncertain etiology of the spasms and the absurd goal of tightening the woman's bust; the humor of the situation dislodges his behavior from genuine erotic instinct, which would be less humorous, and attributes it to an external form of conditioning that has temporarily inhibited his adaptability.[51] This conditioning, an example of Bergson's "mechanical encrusted on the living," makes visible the antithesis of the *freedom* that Bergson sees in the vitality and creativity of human life.

Bergson's most interesting and counterintuitive idea in "Laughter" is to connect automatism with questions of aesthetics and of community that will also be raised by the texts we consider in *Human Programming*.

The social function of laughter, for Bergson, binds communities around the essential vitality of humanity, against which "rigidity, automatism, absentmindedness, and unsociability [stand] inextricably intertwined."[52] For Bergson, "the rigid, the ready-made, the mechanical, in contrast with the supple, the ever-changing and the living, absentmindedness in contrast with free activity, . . . are the defects that laughter singles out and would fain correct."[53] Following Bergson's logic, we laugh at Chaplin's Tramp—at least, when it is his mechanicity that causes our laughter—because we recognize his essential humanity and freedom and want to bring him back within the fold of human community. Could he but hear his audiences through the silver screen, the Tramp would correct his behavior, abandoning his status as an unintentional pervert. In English, "snap out of it!" serves a mechanical metaphor for ceasing to persist in a thoughtless action, and laughter encourages us to snap out of those behaviors by encouraging us to see ourselves from others' perspectives. The comic actor on the stage pretends not to hear the audience's laughter as she muddles about doing something the wrong way, and the comic actor on the screen need not even pretend.

The classic analyses of uncanny automatism follow a surprisingly similar logic, wherein an irrational aesthetic effect resides at the edge of human community, between humanity and objecthood. Indeed, automatic movement is often ambiguously comical or uncanny, depending on how we identify with the automatic entity, and the danger of the situation.[54] The best-known analyses of the uncanny come from Ernst Jentsch's 1906 essay on the psychology of the uncanny and Sigmund Freud's 1919 response. Jentsch, like Freud, discusses E. T. A. Hoffmann's "The Sandman," and he claims that the uncertainty associated with the lifelike doll Olympia—is she real or not?—is the source of the uncanny element in the text.[55] Freud disagrees to the extent that he gives the uncanny a wider, but closely related, significance: "anything that can remind us of [the] inner compulsion to repeat is perceived as uncanny."[56] Freud draws this conclusion from his famous red-light district anecdote, in which the author is lost in a new place and circles back three times, unwittingly, to this disreputable neighborhood. While Freud's eye is on repetition compulsion *(Wiederholungzwang)* as a psychoanalytic concept, he shows us a moment when the body acts outside of the mind's control, making us question our free will or experience the shock of recognizing a repressed desire. This anecdote suggests that moments when we see *our own* automatism, rather than that of another person (like Olympia in "The Sandman"), are the source of a particularly unsettling feeling. Uncanny movements, in these analyses, become uncanny when the body seems to operate entirely outside of conscious control, and comedic ones, as the "mechanical encrusted upon the living," are funny when a mechanical action either perpetuates itself or has an unconscious or alien source. For Bergson, laughter serves the function of trying to put the mechanical behavior to an end and to bring the object of laughter back within the fold of vital community. The sensation of the uncanny serves no such end, and with uncanny others, we often resolve the uncanny's categorical confusions, when we see it in another subject, with a sense of revulsion. Of course, humans are also objects, as anyone who has been to the doctor's office can recognize. Indeed, Foucault's concept of the "medical gaze" in *The Birth of the Clinic* and its continuation through *Discipline and Punish* can make quite clear that those in power treat their populations rather comfortably as objects.[57] If Freud and Jentsch agree, for instance, that epilepsy is uncanny—as a force that takes hold of the body and makes it appear as an object out of conscious control—it's no doubt equally certain that to an experienced physician, epilepsy is not an unsettling phenomenon.[58] When human automatons cause irrational aesthetic effects, they highlight moments when one is uncertain whether to engage in the ordinary coexistence of human acknowledgment (wherein we all behave as though others make free choices as fellow humans) or the very different relationships between humans and objects. We behave differently toward something that has free will, even though we can't see the will directly.

This invisibility and unknowability of free will has been a persistent philosophical problem for centuries, and the human automaton tropes the knotty questions of human agency in a material universe. Observers of a mechanistic universe—made up of atoms or motor neurons or physical entities at another scale—simply cannot locate the will in the material world. Immanuel Kant puts it most succinctly in the following antinomy: when we observe another person, we could reasonably think that his "will is free and yet that it is simultaneously subject to natural necessity, i.e., that it is not free."[59] Depending on our way of perceiving that other person, he may or may not appear more like an object than an autonomous subject. (A version of that question in contemporary analytic philosophy, the "philosophical zombie," asks in which ways we or others might in fact be automatons who only appear to be conscious, as René Descartes and Thomas Huxley famously argued of animals.)[60] Much in the way that the sublime in Kant's account results from a conflict between our abilities to imagine and to understand, we'll treat the aesthetic effects of automatism as the result of our inability to categorize something that appears to be a humanlike object. Reading these analyses of automaton aesthetics alongside Rancière, I find that automatism presents both a negative definition of the human (as the mechanism that makes one appear not-quite-human) and a boundary to an ethical or political community of free individuals. The humorous and uncanny effects of automatism show us moments when the human automaton in question is just out of the bounds of human community and the ways that figuration might be used in negatively defining an idea of freedom, two ideas that I can now consider historically vis-à-vis discouses about race and antitotalitarianism in the United States in the 1930s and 1940s.

Automatons and America

The notion of the human automaton plays an oblique but important role in the history of race and racism in the United States. When we think of the history of race, and particularly the scientific history of race, it is often in terms of eugenics, of the scientific justification of racial hierarchies as reconceived in the late nineteenth century. The paradigm of eugenics and the turn-of-the-century discourse of "atavism," of humanity devolving or turning to a different state, served to describe the races as both biologically determined and definitively ordered in the evolutionary hierarchy. As Walter Benn Michaels and others have traced, thinkers like Lothrop Stoddard and other popular racists offered just such scientific justifications for racist policies of immigration exclusion and nativism in the 1920s and 1930s. Several things changed before and during World War II that brought automatism into the history of racism and propaganda during the postwar period. First, the social sciences, as Kristina Klein has charted, took the baton from biological sciences as the most influential expert method for understanding race. The rise of Franz Boas and a paradigm of antiracist, relativist anthropology, along with the increasing influence of urban sociology in the 1920s, 1930s, and 1940s, provided alternative and often more egalitarian approaches to social policy.[61] Second, the opening of the Nazi concentration camps and the end of World War II irreparably tarnished the reputation of eugenics, along with explicitly racist propaganda that might be linked with it.[62] Finally, as Nikhil Pal Singh and others have observed, particularly with regard to the United States's foreign relations, antiracism was central to the U.S. vision of itself as a world leader (such that the Jim Crow South became a major impediment to U.S. foreign policy).[63] The World War II era thus saw some of the awkward first steps toward color-blind policy and rhetoric as well as the rise of social-scientific research on race, from Gunnar Myrdal's *An American Dilemma*[64] to the 1965 Moynihan Report. This period saw the advent, then, of an official antiracism that, though by no means progressive by today's standards, still differed significantly from the paradigm of eugenic racism.

As we fought World War II and the Cold War, then, explicitly racist propaganda would not do: first because the United States policy makers began to describe democracy as antiracist, second because our enemies had been racist, and third because our Cold War enemies, including our erstwhile allies the USSR and China, became increasingly difficult to define in strictly racial terms; the enemies became, more than just enemy nationalities, the ideologies of fascism and communism, along with the control we began to call "totalitarianism." As we see throughout *Human Programming*, texts that make use of automaton imagery often purport to describe subjects independently of biologically essentialist racism. Nevertheless, that biological essentialism often reemerges in automaton discourse in surprising ways, to which we'll pay close attention: sometimes it will emerge as a paternalistic attitude in the midst of an attempt not to be racist, and other times it will appear in descriptions of specific cultural groups, particularly Asians and Muslims, as unusually prone to automatism. If the biology of the era of eugenics and atavism emphasized in a Darwinian sense humanity's resemblances to animals, then automatism would emphasize humanity's resemblance to machines—and, pointedly, machines whose workings might be manipulated by experts. So we see in the period two ways of being less than human: as a human animal and as a human machine. If atavism represents a descent into the body and into instinct, into a primitive nature, then automatism would emphasize an explicitly *unnatural* state of affairs brought about by industrial and commercial modernity, Oriental hypnosis, voodoo magic, cult programming, or brainwashing. This unnatural automatism could be represented with particular ease on film as well as in prose descriptions of automatic movement, apparent programmability, and uniform mass consciousness.

Postwar descriptions of humans' resemblance to machines do have precedents, however, in the early twentieth century. Colleen Lye, for instance, has analyzed how instigators of the "Yellow Peril" in the first decades of the century, including Jack London, proffered images of supposedly machinelike Chinese workmen who were hired to build railroads in the western states. Lye offers, relative to widespread comparisons of Asian Americans to machines, a description of the "modern" racial essence of the Asian American whose echoes I will trace later in the century: "the brute" of atavistic racism, Lye writes, "is typically a kind of 'wild man,' desire incarnate loosed from social control, denoting the figure of primitivism within modernity. The coolie signifies a different kind of monstrous presence, not the ambivalent pleasure of the body's libidinal release, but, on the contrary, the prospect of its mechanical abstraction."[65] This kind of comparison to machines and mechanical abstraction would spread to many other groups through the course of the twentieth century, and the example of the Haitian zombie can allow us to see more clearly the difference between humans who resemble animals and those who resemble machines.

The zombie was introduced to U.S. culture by the travel writer William Seabrook during the U.S. Occupation of Haiti, which lasted from 1915 to 1934.[66] In his popular Haitian travelogue *The Magic Island*, Seabrook recounts in one chapter how he first saw came face-to-face with zombies: he encounters three workers who seem to be dumb brutes, and when one stands up, "like an animal," "the whole face . . . was vacant, as if there was nothing behind it."[67] Seabrook experiences a textbook Freudian uncanny moment, noting that the sight "'upsets everything' . . . [all] the natural fixed laws and processes on which all modern human thought and actions are based."[68] Gaining control of himself, he is struck with a recollection from the "histological [tissue study] laboratory at Columbia."[69] In a conceptual move that Seabrook repeats later in the narrative, he invokes a familiar space—Columbia University in New York City—to diminish the strangeness of what he sees in Haiti. This memory from Columbia is that of the "face of a dog" whose "entire front brain had been removed in an experimental operation. . . . It moved about, it was alive, but its eyes were like the eyes I now saw staring."[70] The experimental operation he refers to is a prefrontal lobotomy, a technique that would not be practiced on human patients in the United States until 1936, by Walter Freeman and James Watts. The botched lobotomy, the image of which has become most familiar through the final scenes of Milos Forman's film *One Flew over the Cuckoo's Nest* (1975), is the image that Seabrook wants to conjure. While the zombie's resemblance to an animal might in the first place suggest an instinctual, savage being, Seabrook turns instead to an image of detachment. Resembling a dog and resembling a *lobotomized* dog turn out to be quite different things. The eyes of the supposed zombie, which had been "like the eyes of a dead man, not blind, but staring, unfocused, unseeing,"[71] are disconnected not only from emotional connection to others but even from the sense of sight itself. Such is the form of detachment that the zombie, in this iteration from the 1920s and 1930s at least, embodies. This image of the zombie was quite distinct from an example of bestial racial essence from just a decade earlier, in Eugene O'Neill's 1919 play *The Emperor Jones*, where Haiti sets the scene for a black man's atavistic devolution into a beast.[72] In Seabrook's story, the threat posed by the other has less to do with racial essence than with an exoticized technology.

Three years after the publication of *The Magic Island,* the film *White Zombie* features a factory that elaborates on this unnatural form of subhumanity within the visual field. Although *White Zombie* contains few formal innovations beyond the early horror films that preceded it—

Dracula (Tod Browning, 1931), *Frankenstein* (James Whale, 1931), *Metropolis*, and *The Cabinet of Dr. Caligari* constitute its playbook—this Halperin brothers film succeeds in investing the uncanny effects of cinematic automatism with the rich context of Haiti's U.S. takeover and industrialization. In the film's most famous scene, the protagonist Beaumont takes a tour of a factory run by Murder Legendre, played by Bela Lugosi on the heels of his performance as Dracula. The introduction of Legendre follows a long exposition of the factory's workings: first, the zombies walk in file on an elevated walkway, carrying baskets of sugarcane on their heads in a manner that resembles a human conveyor belt. Next, one of these workers seems to misstep in the spot where the workers dump their cane into the thresher, but his body remains rigid and falls into the thresher. The qualities of movement in the zombie's fall help to underscore the fact that these are indeed zombies and not simply well-disciplined workers. The macabre effect of this scene is amplified when the camera moves from the immediate scene of the grinding blades toward the men at the bottom who continue, unflinching, to turn the wheel. The groan of the machine continues as the film cuts to a medium shot of the zombies' faces as they turn the wheel, with an expressionless and no doubt redundant zombie overseer in the left background. What's shocking about these subhuman creatures is their lack of emotional response to each other's peril, as they have been programmed to perform their labor to the utmost efficiency. The Halperin brothers introduce in this film a fantasy of subhuman machine-people between whom there can be no community, no ethical bonds, and nothing like democracy. The overseer in the background watches on with folded arms, doing nothing. Unable to preserve their own (half-)lives or each other's, the zombies' uncanny quality springs from the nonresponsive eyes and faces that accompany their bodies' continued exertions.

This kind of representation caught on and was quickly used in horror film depictions of other non-American populations. The second zombie film by the Halperin brothers, *Revolt of the Zombies* (1936), tested the zombie image's portability across global contexts. In this similarly low-budget picture, the myth of the zombie moves to World War I–era Cambodia, where the zombie is the product of a curse on the Cambodian city

The factory in Victor Halperin's *White Zombie* (1932), where Beaumont (Robert Frazer, at left) looks on in horror as Haitian zombies mindlessly turn the wheel that grinds their fallen comrade in the sugar mill.

of Angkor, a backstory that is wholly unrelated to Haitian voodoo. The ancient city of Angkor, in the film, had been built by "robots," "automatons," or "zombies"—terms used interchangeably until science fiction about robots became popular[73]—and some of these zombie soldiers of the same variety have been brought to the Franco-Austrian front. The most striking image in this film would be the marching onslaught of zombie soldiers, whom we see advancing toward the enemy even after they have been shot through the heart. This relocation of the zombie trades in the suggestive power of exotic religion, even as it connects the notion of human automatons to a longer tradition of Orientalist notions of hypnotism, opium use, reincarnation, and so forth. *Revolt of the Zombies* borrows, then, not only from *White Zombie* and *Dracula* but also from *The Mask of Fu Manchu*, in which a yellow-faced Boris Karloff wields magically hypnotic powers. This iteration of the human automaton, then, could be adapted to represent various demographics, independently of the racial stereotypes that could also be employed.

Thus, by the middle of World War II, a representational paradigm for the new problem of "totalitarianism" was ready to hand for anyone who wanted to borrow from depictions of factory automatism, as in *Modern Times* and *Metropolis,* or horror images of the automaton as unnatural and uncanny other. As propaganda for the war effort struggled with racial types with the multinational enemy, there was some uncertainty about how to represent fascism. Chaplin's own *The Great Dictator* (1940) eschewed automatism for a cast of comical thugs and buffoons, but it was a film that did not face the challenge of representing Japan alongside Germany and Italy. The mass, automatic movement of marching soldiers became one of the signature visual motifs with the beginning of Frank Capra's Why We Fight series for the U.S. government (1942–45). The series of magnificently produced propaganda films, shown to soldiers before heading to battle, made a strong case in the first film, *Prelude to War,* for the distinctions between the "Free World" and the "Slave World."[74] One of the film's most prominent repeated motifs takes images of marching masses of soldiers and overlays them onto maps, literally creating a global mapping of "us" and "them," free American selves and totalitarian automatons.

During the war, Walt Disney also became involved in the wartime propaganda effort, producing approximately a dozen films for the U.S. government.[75] One of Disney's visual representations of the totalitarian automaton took the form of an explicit update of *Modern Times,* in the Donald Duck cartoon *Der Führer's Face* (1942). There, Donald experiences a day-in-the-life of Nazi Germany, working on a munitions conveyor belt in a clear homage to Chaplin's factory. He experiences the frenzy of the accelerating assembly-line belt, but the Disney animators adapt it humorously to the new situation by including portraits of Hitler interspersed among the munitions on the belt. The automatic actions of tightening the munitions trade off in Donald's routine with the automatic, compulsory action of saluting each approaching picture with the right arm, "Heil Hitler." The motion is initially comical, making light of the mechanical behaviors of the factory and of the totalitarian crowd,

A map overlay of Italian soldiers marching and saluting, part of the display of the totalitarian "Slave World" in Frank Capra's *Prelude to War*, the first newsreel of the Why We Fight U.S. wartime propaganda series (1942–45). Capra makes use of the heil salute and long shots depicting marching soldiers and crowds to depict mindless totalitarian masses in these films.

but as this bolt-tightening-and-saluting sequence accelerates and continues, the comedy of *Modern Times* becomes a horrific onslaught of machines and Führers. That transition takes shape in the Disney animation through swirling, hallucinatory specters of Hitler, belching work bells, and missiles—and, of course, the panting, desperate cartoon face of a bewildered Donald Duck. Donald is trapped within the factory and he cannot escape, until, luckily, he wakes to find that his life in the totalitarian state had been a mere nightmare.

The most chilling of the Disney propaganda films produced during World War II is titled *Education for Death* (1943); it follows the story, from birth to adulthood, of a boy named Hans born in Nazi Germany. The animated short's final sequence carries us from Hans's charming boyhood to his villainous adulthood, a transition achieved through a long growth sequence that shows him continually marching as he grows older. "Heiling and marching, marching and heiling," Hans marches as the narration describes how "the grim years of regimentation have done their work," turning Hans into a "good Nazi." He sees, says, and does only what the Nazis want him to: the cartoon's additions of swastika horse blinders, a muzzle, and a chain around the neck enliven this ever-marching image of limited vision, speech, and action within the Nazi regime. *Education for Death* studies the *production* of the enemy, as the innocent young Hans had seemed at first as though he would become an ordinary child. Instead, the "marching and heiling" and schooling produce an enemy directly out of the fascist education system, which has made him an ethically reprehensible figure into whom "no seed of laughter, hope, tolerance, or mercy" has been planted. Rather than through Orientalized or primitive magic, Hans's unnatural automatism has been effected through the use of propaganda itself.

There develops across these and other World War II propaganda films an iconography of marching as empty motion, of fascist consciousness as an empty, instrumental soldiering. Of course, American soldiers marched, too, as all soldiers must, a fact that might make the fascist marching seem like an odd representational choice. Leni Riefenstahl's *Triumph of the Will* (1935) featured many shots of crowds saluting Hitler, soldiers marching in unison, and Hitler Youth playing instruments, as images of sublime force and solidarity, although it also often features shots of individual audience members listening to and comprehending the Party speeches. The ex-Nazi Hermann Rauschning would write in 1939 that "the simplest . . . most effective and most characteristic method of domination employed by National Socialism . . . [is] the marching."[76] Though this may seem like a curious judgment, it attests to the fact that marching is the most visible signifier of the fascist crowd's mass consciousness and of the totalitarian government's control. Rauschning thus calls marching "an indispensible magic stroke performed in order to accustom the people to a mechanical, quasi-ritualistic activity until it becomes second nature," a description that can conjure some of the aesthetic power of the image.[77] Riefenstahl's and others' images of marching, then, provided Capra and the Disney filmmakers with a set of visual, cinematic cues that signify the mass, automatic consciousness of the Nazi soldier in what Capra calls the "Slave World." The automatic heiling arm appears to be a corollary to this automaton logic, and it, too, would be identified with Nazi consciousness, as a sort of slavishly obedient unconsciousness. With this image, Peter Sellers's performance as Dr. Strangelove in Stanley Kubrick's 1964 film of the same name could play the right, "heiling" arm as a pantomime of a Nazi consciousness, the undying conditioned belief and motion that, even after the end of World War II, could not be exorcised. In a situation where racial stereotype was not helpful to propagandists seeking to locate the otherness of the enemy, their movements and their minds—rather than the racial characteristics of their bodies—became the focal point of these representations. Likewise, by the same paradox of freedom's invisibility, the best way to represent American freedom was through the unfreedom of an automaton totalitarian consciousness.

Following World War II, the image of automatism would be increasingly attached in the United States to the question of what it means to

Peter Sellers as the title character of Stanley Kubrick's *Dr. Strangelove, or: How I Learned to Stop Worrying and Love the Bomb* (1964). Sellers plays on the well-known iconography of the Nazi heil as a comic gag wherein his arm has been conditioned independently of his body.

be free or unfree. The automaton's visible unfreedom resolves a formal problem in literature, film, and political theory, because freedom in the abstract, as Kant argued, cannot be seen directly. Moreover, in a paradox that Kant develops in the passage from his first *Critique* to his second (from epistemology to ethics), freedom is *both* entirely invisible *and* a necessary precondition to ethics, which must admit some possibility of free choice.[78] And the same may be said of American democracy, because nominally free choice and informed self-interest are cornerstones of a purportedly self-ruled people. The historical meanings of American "freedom" have been multifaceted over the centuries, from the Puritans' City on a Hill (religious freedom) to Frederick Jackson Turner's "frontier thesis" (the freedom of geographic expansion) to the American civil rights movement's "freedom riders" (freedom for all to exercise civil rights) and the Tea Party's revivals of the "Don't Tread on Me" flag (freedom from governance). In January 1941, nearly a year before the United States would declare war, President Roosevelt articulated in his "four freedoms" State of the Union Address the freedoms of speech and of worship, freedom from want and freedom from fear, as those the government must protect. The third, freedom from want, referred to trade policy but would also have been in keeping with the Keynesian principles of the New Deal, and the fourth freedom, from fear, resulted in a call for the worldwide "reduction of armaments."[79] The *Saturday Evening Post* illustrator Norman Rockwell would go on to make illustrations of each of these freedoms. Meanwhile, the Chicago economist Friedrich Hayek was writing in *The Road to Serfdom*, in lines of thought he would reiterate in 1960's *The Constitution of Liberty*, that state control of the economy in general was totalitarian and that "the economic freedom which is the prerequisite of any other freedom cannot be the freedom from economic care [want] which the socialists promise us. . . . It must be the freedom of our economic activity which, with the right of choice, inevitably also carries the risk and the responsibility of that right."[80] Hayek and other liberal and neoliberal economists would present an equal and opposite notion of freedom from FDR's "freedom from want," and theirs would also take advantage of antitotalitarian rhetoric and the specter of humanity shorn of the fundamental right of choice, the "prerequisite of any other freedom." Looking toward the present, the ideal of freedom has also underpinned xenophobic agendas, and the specter of a simultaneously fundamentalist and totalitarian unfreedom hangs in right-wing discussions of the possibility that Muslims, or even Barack Obama, might impose Sharia law in the United States. American conservatives often invoke totalitarian unfreedom in describing even political correctness, as David Horowitz and Robert Spencer did when they borrowed a term from Orwell to title a recent book *Islamophobia: Thought Crime of the Totalitarian Future*.[81] Because freedom is a flexible and capacious idea that has been recruited into many political agendas, *Human Programming* focuses on ways that many of the most abstract representations of American freedom have depended on images of sublime unfreedom for contrast.[82] As Americans contested the meanings of freedom and American-ness, in the face of shadowy enemies of democracy such as totalitarians, extremists, and fundamentalists, cinematic and novelistic images of automatism were on hand as means of representing the difference between Capra's "Free World" and the "Slave World." The pages of *TIME* magazine, science fiction film and television, anti-totalitarian psychological studies, and even courtroom testimony about cults would take up the task of describing that "Slave World" and its denizens through several iterations over the following decades. This set of strategies for describing unfreedom produces not just its mirror image of American freedom but also its own, peculiarly American, vision of unfreedom.

Insofar as descriptions of unfreedom stand as negative definitions of freedom itself, *Human Programming* situates these depictions of totalitarians, communists, cult members, extremists, and terrorists as a major current of what scholars including Sacvan Bercovitch and Donald Pease have described as discourses of "American exceptionalism." In *The New American Exceptionalism* and elsewhere, Pease describes the interdisciplinary footprint of twentieth- and twenty-first-century American exceptionalism, a set of ideals that defined "an academic discourse, a political doctrine, and a regulatory ideal assigned responsibility for defining, supporting, and developing the U.S. national identity."[83] According to Pease, "the exceptionalist paradigm described U.S. uniqueness in terms of what the nation lacked," particularly in terms of its social formations: "a landed aristocracy, a feudal monarchy, a territorial empire, a society hierarchized by class, a deeply anchored socialist tradition."[84] That double negative—defining freedom by describing it as lack of unfreedom—is also a key element of the more nightmarish images of the human automaton, whose enactments of mental unfreedom render abstract freedom visible. The material circulation of these images of automatic, unfree, and therefore subhuman enemies thus makes it possible to trace both inherently unstable notions of an exceptional American freedom and the shifting margins of an increasingly globalized American community. This method owes much, too, to other American studies scholars who have emphasized, in various ways, the aesthetic and affective dimensions of the public sphere.[85]

Throughout my own examination of the aesthetics of unfreedom in the U.S. public sphere, I pay attention to the ways that scientific paradigms have shaped or been used to shape our perceptions of others. Nevertheless, I build on an approach to the relationship between humanistic and scientific approaches that does not view these proverbial "two cultures" as antagonistic or even quite separate.[86] Rather, I want to suggest that tracing networks of exchange between literature and science can help us to avoid the limitation of imagining literature as essentially human and humane in its concerns, while denigrating science as essentially inhuman, dehumanizing, and reductive. In fact, chapter 2 of *Human Programming* aims to contribute to a genealogy of this very notion by charting the popular image of scientific institutions as sites of sublime discipline and unfreedom, an image that developed through direct literary borrowings from the antitotalitarian discourses of the 1940s and 1950s. Along these lines, I acknowledge the critical impulses and ethical imperatives put forward in novels and films, at the same time as I insist that their most important political consequences lie in the circulation of their images. What I find in the course of this approach is one of the most widespread, as well as rhetorically and aesthetically powerful, literary and cinematic devices in postwar U.S. culture. Using the human automaton in literature and elsewhere, Americans have drawn political conclusions from psychological premises, and vice versa, in a near-continual contestation of the meanings of freedom and unfreedom.

From within this framework, the chapters of *Human Programming* chart how the automaton has spread throughout U.S. culture and discourse in the post–World War II period. The first chapter, "Uniquely American Symptoms," addresses the prospect of the automatic mind that developed out of propagandistic uses of the human automaton between World War II and the decade following the Korean War. This chapter traces the spread of ideas of automatic consciousness, propaganda, and what I call "totalitarian language," from Orwell's *1984* to the development of the Cold War "brainwashing" scare and the masterful satires of *Shock Corridor* (Samuel Fuller, 1963) and *The Manchurian Candidate*. Within the brainwashing discourse, I discern a dialectic between a free democracy and an unfree, totalitarian other that would define public-sphere uses of the human automaton from early World War II through the War on Terror.

Human Programming's second chapter, "Anti-institutional Automatons," charts out how progressive figures in postwar literature and politics reappropriated the automaton image to express new ideas about individual freedom and institutional unfreedom in the 1950s and 1960s. Ralph Ellison's innovative and influential use of automatons in *Invisible Man*[87] provided a template for other writers to compare the technocrats of the mid-century American establishment to the masterminds of totalitarian states. This influential reversal of the automaton-as-other was a key element of the progressive image of individuals rebelling against institutional enslavement, which I trace through the work of Ken Kesey and Betty Friedan, among others. These writers developed the human automaton image by focusing on the social and institutional conditions that they thought curtailed mental freedom and spontaneity; moreover, they reframed political problems like racism, exploitation, and patriarchy as "total institutions" that reduced their inmates to automatons. As such, the image of the automaton developed as a primary vehicle for discussing abstract conceptions of individual freedom for conservative Cold Warriors, for anti-institutional progressives, and even, as I describe at the end of the chapter, for commercial culture, libertarianism, and Silicon Valley cool.

Chapter 3, "Human Programming," retraces the historical steps of the first chapters, from the 1940s to the 1970s, alongside the cutting-edge technologies of computation, genetics, and cybernetics and their ripples throughout science fiction through the end of the Cold War. Beginning in the 1940s, but stretching into the present, the intertwined developments of cybernetics and the postindustrial economy inspired additional transformations and refinements of the automaton image. Whereas the 1930s industrial image of the human automaton emphasized programmed movements and statements, in the wake of cybernetics, this image could reflect the devaluation of physical and, increasingly, intellectual and affective labor in the postindustrial economy. In response to these problems, writers including Kurt Vonnegut, Philip K. Dick, William S. Burroughs, Shulamith Firestone, and Neal Stephenson explored the automatic dimensions of the apparently human domains of emotion and belief. Even on this shifting technological terrain, the human automaton figure proves amenable to representations of human selves and subhuman others, in new adaptations of Orientalism and in new representations of fundamentalism. Complementing posthumanist scholarship that imagines a human past and posthuman future, in physical or philosophical terms, I suggest that influential science fiction like Stephenson's *Snow Crash*[88] advances a darker perspective on the posthuman: a geopolitical mapping of free, human selves defined in ongoing opposition to posthuman, unfree others.

The final two chapters of *Human Programming* trace how the human automaton has spread since the 1970s in depictions of radicals, cult members, and fundamentalists. Chapter 4, "Cult Programming," explains how what I call the *topos* of the cult developed out of psychiatric research on Korean War POWs, which described charismatic leadership, social isolation, and specialized language as criteria that delimited the social space of mental unfreedom. Between the 1970s and the 1990s, the cult became the subject of U.S. news media moral panics, young adult literature, popular television, and postmodern literary fiction. Literary conventions and Cold War psychology shaped the ways that experts have described extremist groups, from Patty Hearst's Symbionese Liberation Army to the Unification Church (1970s to the present) and the mass suicides of Jim Jones's People's Temple (1978) and Heaven's Gate (1997).

Chapter 5, "Fundamentalist Automatons," returns to the moment with which I began this introduction, the War on Terror, and describes the roles automatism has played in discourses of Islamophobia since 9/11. The high culture of the War on Terror, including serial television, documentary, and literary fiction, is frequently structured around the notion of a pathological terrorist consciousness, borrowed from the currents of thought I describe in the previous chapters. This most recent iteration of the human automaton figure in science fiction attempts to describe complex forms of compulsion and automatism within free subjects, but with frequent reference to "brainwashing" plots and the tropes of robotics, as in the television series *Battlestar Galactica*. Likewise, realist texts, such as Showtime's *Homeland* and Don DeLillo's *Falling Man*,[89] use the automatisms we associate with popular diagnoses and pathologies, such as the repetition compulsion of post-traumatic stress disorder (PTSD), as narrative means for comparing the consciousness of the terrorist with that of the ordinary American. This unusual literary deployment of the language of diagnosis and pathology allows these texts to put pressure on the Cold War–era dialectic of free selves and unfree others, as it encourages readers to understand terrorist psychology and ordinary Americans' psychology on the same footing.

Even as *Human Programming* traces how individual agency has been represented and retooled in the United States since World War II, it also reveals some of the ways that our critical understandings of agency and determinism might be reframed altogether. I turn in the conclusion to a science fictional description of the programmed mind in Daniel Suarez's 2006 novel *Daemon*,[90] a text that uses automaton imagery in the course of exploring network forms as an alternative schema for understanding individual and collective action. Despite the ways it might limit our thinking, the image of the human automaton nonetheless continues to shape the way we describe freedom and unfreedom in the U.S. public sphere. We can try here to understand its aesthetic and political power.

CHAPTER ONE

Uniquely American Symptoms

*Cold War Brainwashing and
American Exceptionalism*

E arly in the 1962 film *The Manchurian Candidate*, the master brain-washer Yen Lo describes the benefits of brainwashing, as he shows his audience his sleeper assassin, Raymond Shaw. Through the process of brainwashing, Dr. Yen explains, the assassin has been "relieved of those uniquely American symptoms of guilt and fear."[1] As a result, there will be "no possibility of his being caught." One of many brilliantly executed moments of satire in the film, Dr. Yen boasts that his scientific knowl-edge is advanced enough to have located—and eliminated—these main weaknesses of the American personality. It is no accident that such weak-nesses would be "guilt and fear," traits that not only supposedly make one American but, more important, make us human. How human can one be, the film asks, without guilt, the basic mechanism of moral account-ability, and fear, the core of self-interest and self-preservation? By play-fully mentioning these "uniquely American symptoms," the film takes aim at an ideology of American exceptionalism that developed within a variety antitotalitarian writings of the 1940s and 1950s. Although there was much variation between antitotalitarian of sociological, psychologi-cal, and philosophical expertise in the mid-century period, many of the thinkers associated with it described the contours of totalitarian psychol-ogy through contrasts with American humanity and freedom. Take, for instance Joost Meerloo, the Columbia University psychiatry professor who wrote in 1956 that, "in the totalitarian countries, where belief in the

Pavlovian strategy has assumed grotesque proportions, the self-thinking, subjective man has disappeared."[2] A decade before the death of man heralded by Michel Foucault's *The Order of Things*,[3] Meerloo finds the death of the ethical subject in the totalitarian state.[4] Even though behaviorist research was alive and well in the United States, particularly in B. F. Skinner's utopian and often visionary writing, the negative consequences of behaviorism, technocracy, and propaganda were often attributed, as in Meerloo's case, to Ivan Pavlov and his followers. The "brainwashed" totalitarian subject, as a fully programmable bundle of habits, represents for these thinkers the death of the liberal subject, along with the subjects of democracy and of ethical community. When these thinkers imagine the ways to combat the problem of brainwashing, they articulate a prominent logic of American exceptionalism.

While the images of the totalitarian automaton I mentioned in the introduction were defined by automatic movement, the 1940s and 1950s also saw the development of automatic *language* as a defining characteristic of the totalitarian parts of the world. After World War II, the possibility that totalitarian countries might use advanced propaganda technologies to manipulate minds loomed large in the American imagination. This problem of programmed minds and unfree subjects began in the antitotalitarian writings of the 1940s, flourished with the coinage of the term *brainwashing* in the early 1950s, and then developed further through the brainwashing texts, fiction, and films of the 1950s and early 1960s. Because it proved so useful for describing a nondemocratic enemy, we can take the figure of the brainwashed human automaton as a central trope in U.S. Cold War culture and also as one of that culture's enduring legacies in the present.[5]

Totalitarian Language

During the Cold War, a new conception of unfreedom emerged across U.S. culture, from science fiction pulps to political theory, one that placed American freedom in a dialectical relation to totalitarianism and its techniques of domination. Analyses of machine-age alienation—such as William Seabrook's 1931 image of modern man as a "cog in a wheel"—

had provided a shared grounding for the early cultural analyses of totalitarianism, many of which were undertaken by German émigrés associated with the Frankfurt School.[6] Theodor Adorno's mid-century writings, for instance, advance the claim that capitalist modernity and fascism are two sides of the same coin, that is, that the one face of the "administered world" gives way readily and easily to the other.[7] Adorno's Frankfurt School colleague Erich Fromm was much more widely read in the United States during the period, however, and his 1941 *Escape from Freedom*[8] became a long-term best seller in the United States, to be issued in a second edition in 1965. Associated with Columbia University after his emigration from Germany, Fromm combined psychoanalysis and Marxian analyses of alienation to develop an account of totalitarianism as a fundamentally psychological problem. The whole of *Escape from Freedom* is arguably an expansion of an insight made by John Dewey in the 1939 book *Freedom and Culture,* which Fromm cites at the outset:

> The serious threat to our democracy . . . is not the existence of foreign totalitarian states. It is the existence within our own personal attitudes and within our own institutions of conditions which have given a victory to external authority, discipline, uniformity and dependence upon The Leader in foreign countries. The battlefield is also accordingly here—within ourselves and our institutions.[9]

Fromm's approach to the problem of totalitarianism, borrowing from and expanding on this direction from Dewey, takes the complex workings of international politics out of the picture, at the same time that it gives the ordinary reader something to do. The project of examining our attitudes and our national psyche, rather than critiquing firmly embedded institutions—as Adorno was doing in the mid-1940s with respect to the culture industry—would take hold in many Cold War writings as a way to seek out a moral, human essence of democracy that might be celebrated as the key to American freedom.

Escape from Freedom, then, treats Nazism as a "psychological problem" and sets about differentiating the personality traits that embrace freedom from those associated with the fear of freedom.[10] First, it is industrial modernity that provokes feelings of alienation, providing the temptation to conform, to "give up" one's "individual self and [become] an automaton, identical with millions of other automatons around" one.[11] Moreover, Fromm will contend that we might not recognize our own false consciousness as such. Fromm uses the example of the hypnotic experiment as a way to describe the false consciousness he associates with human automatism: "although one may be convinced of the spontaneity of one's mental acts, they actually result from the influence of a person other than oneself under the conditions of a particular situation."[12] Many decisions, then, are even unwittingly driven by a "fear of isolation" that is part of the alienation of modern life.[13] The task of rooting out false consciousness and programmed attitudes could thus become an endless one. In a conclusion drawn from 1930s depth psychology, Fromm asserts that we must defend against conformity and false consciousness (the "pseudo self") through the "realization of . . . self" and the exercise of spontaneity.[14] For Fromm, "artists" are the chief exemplars of the spontaneity to which Americans could aspire, as the torchbearers of uncoerced enjoyment and creativity.[15] Overall, that solution seems less compelling than the problem that he articulates. He identifies the possibility that "we have become automatons who live under the illusion of being self-willing individuals" and the connection between a psychological disposition toward conformity and the politics of totalitarianism.[16] Strategies for ensuring that one was not creeping toward a totalitarian automatism in one's daily life would become the subject of the propaganda effort on "brainwashing" throughout the 1950s, which I discuss in the following section.

Fromm's approach to the problem of totalitarianism shared much in common with critiques of propaganda, despite the structural emphasis of the latter. The development of the media technologies enabled propaganda during and after World War I, and they became more and more sophisticated in the decade leading up to World War II. Mark Wollaeger has traced how the word *propaganda* in English usage, a synonym for *information,* had no explicitly negative connotations until the late 1920s, since which time propaganda has been the object of near-constant critique

and discussion.[17] The persuasive techniques available to camera operators in cinematic propaganda—Capra's Why We Fight series is a verita-
ble catalog of brilliantly affecting montage work—met with audiences
that virtually all the early theorists of the cinema assumed to be utterly
uncritical. And particularly in the wake of Adorno's writings on the cinema and advertising as a tightly orchestrated culture industry, it is un-
surprising that propaganda has often been considered as a problem that
closely follows developments in media technology. Meerloo, one of the
authors who wrote extensively on brainwashing in the 1950s, saw the
"creeping coercion by technology," especially television, as a dangerous
medium, by virtue of its "hypnotizing, seductive action."[18] In Meerloo's
view, media developments would lead, not to a third world war, but to a
"tremendous battle between technology and psychology . . . of systematic conditioning versus creative spontaneity."[19] In the past two decades,
critics, including Friedrich Kittler, Stefan Andriopoulos, and Alan Nadel,
have posited that the very forms of television and film are suggestively
similar to hypnosis and brainwashing.[20] Andriopoulos and Kittler seize
on the older writings of Raymond Bellour and Jean Cocteau, respectively,
which both suggest, à la A Clockwork Orange (Stanley Kubrick, 1971),
that the scene of the theater as a closed space of reeducation can be considered as akin to hypnosis. Nadel has argued that open and universal
simultaneity of broadcast television is a necessary condition for Americans to imagine the kind of mass hypnosis that Edward Hunter and
others would claim had occurred through other means in China.

George Orwell, the British novelist who held a uniquely widespread
influence in postwar U.S. thought and culture, took this media-driven
argument about propaganda and human programming in an innovative
and enduring direction. He asked what happens if we consider language
itself, rather than the radio, the newspaper, or the cinema, as the ultimate medium for propaganda. The novel 1984 features the "two-minutes'
hate," which elicits and shapes raw negative emotion in the cinematic
mode, while the newspapers that Winston Smith rewrites for the Party
primarily control the distribution of information in print. But more important to the novel than the propaganda powers of the press or the cinema is the ruling Party's treatment of language itself as a manipulable

channel of thought. An enthusiast for early variations on Esperanto, Orwell had a serious and long-standing interest in programmatic language reform, a signal feature of his well-known essay "Politics and the English Language" and of *1984*'s appendix "The Principles of Newspeak." He fought to keep the latter, at some financial risk, in the American Book-of-the-Month Club edition of *1984*.[21]

Through these writings, Orwell develops a popular theory of what I call here "totalitarian language" that would prove influential to Hannah Arendt, Edward Hunter, and the many others who have taken on the assumption, inherent in "Newspeak," that censorship's limitations on the press and public speech could be echoed on the levels of language and thought.[22] In "The Principles of Newspeak," Orwell offers the example of "Comintern" as the principal real-world reference point for the novel's patterns of portmanteau Newspeak (e.g., "Minipax" for Ministry of Peace, "thinkpol" for Thought Police).[23] There is for Orwell a significant difference between the portmanteau and the original: "*Comintern . . .* suggests merely a tightly knit organization and a well-defined body of doctrine, a word that can be uttered almost without taking thought. . . . *Communist International,*" conversely, is, according to Orwell, a "phrase over which one is obliged to linger at least momentarily."[24] The media technology of the portmanteau, in Orwell's thinking here, modulates the speed of thought and reflection. Orwell thus inverts Victor Shklovsky's theory of artistic defamiliarization—in which art draws prolonged attention to ordinary objects we usually pass by thoughtlessly—positing, rather, that speeding up perception might prevent the speaker from having any thoughts whatsoever. Just as Shklovsky posited that the amplified descriptions in great literature could encourage the reader to see things in a new way, Orwell imagines the equal and opposite media technique as it operates within official language.[25] And he pursues the political consequences of language's speed outside of *1984*'s Oceania, as well, particularly in the essay "Politics and the English Language," an appeal in part to eliminate set phrases and clichés from writing. While he makes other appeals to good writing, it is through the practice of using set phrases and tired metaphors without stopping to visualize their meanings—for example, "the fascist octopus has sung its swan song"—that bad thinking can be seen within bad writing.[26] "If one gets rid of these habits," Orwell writes, "one can think more clearly, and to think clearly is a necessary first step toward political regeneration."[27] That "political regeneration" is both serious and implicitly democratic, and I read Orwell's turn to language as a search for the thoughtfulness and reflection that might stave off the mindlessness he associates with the totalitarian subject. In *1984*, that totalitarian mindlessness is exemplified in the delightfully named "duckspeak," which denotes the style of enunciation that is Newspeak's end goal. Winston hears a phrase "jerked out very rapidly" in such a way that he is certain "that every word of it was pure orthodoxy, pure Ingsoc."[28] Suggesting William James's notion of habit as the intelligence that takes root in the body without involving the higher brain centers, Winston imagines the duckspeaker as "some kind of dummy" because "it was not the man's brain that was speaking, it was his larynx."[29] Winston's disdain for the duckspeakers around him constitutes one of his first rebellions in the novel, a rebellion against the plan for a perfected totalitarian language.

This idea of Orwell's was anything but an isolated case. Consider Hannah Arendt's descriptions of Adolf Eichmann's personality, which she tied to several descriptions of his language. When Eichmann declares in the trial that "officialese [*Amtssprache*] is my only language," Arendt takes this admission as a profound indication of his condition: "what he said was always the same. . . . His inability to speak was closely connected with his inability to think, namely, to think from the standpoint of somebody else."[30] Arendt here points to the impossibility of empathy as the result of a life within "officialese," but her statement also bears much in common with Lionel Trilling's conception for critical thought, the liberal imagination that was alone capable of resisting and thinking beyond the confines of propaganda and dogma. From a more politically conservative standpoint, Ayn Rand had also experimented with a similar relationship between linguistic form and personal knowledge in her early novella *Anthem*,[31] wherein the future dystopia has banished individualism along with the first-person singular pronoun. Orwell's and his contemporaries' ideas about language may appear as quaint relics after deconstruction, as his "Comintern" example presumes the ability to detect either the presence or absence of thought itself within language. This vision of language is also one in which human creativity is unable to flourish within a restricted language, in a direct analogy to the way that the literature of dissent is unable to flourish with state censorship.[32] In imagining language as a microtechnology for manipulating thought, Orwell and his contemporaries provide a missing link between behaviorist theories of the subject, in which habits might be changed based on stimulus and response, and theories about the workings of propaganda, which can broadly disseminate misinformation. And the plot of *1984* builds this bridge between behaviorist conditioning and propaganda even more thoroughly.

Orwell's thinking is suffused with the assumptions of behaviorist psychology, because his theory of propaganda's human programming was grounded in habit. Orwell borrows his central torture scene, in which Winston is brought face-to-face with rats, from the behaviorist psychologist John B. Watson's "Little Albert" experiment. But perhaps the more profound debt to Watson in *1984* is the treatment of language as a set of habits—habits and nothing more. By eschewing theories of consciousness altogether, Watson and his followers had arrived at peculiar but important theories of mind and of language: theirs is a model of the human being in which interior states were necessarily unimportant. Watson develops a theory of language wherein (1) "words are substitutes for objects" and (2) words are "equivalent to objects in releasing behavior."[33] According to Watson, "the behaviorist's theory of thought hinges upon the way word habits are formed—upon word *conditioning*." The most radical consequence of such a view is the possibility that the subject's most cherished values and ideals are, at bottom, *habits* of thought. And in his "The Principles of Newspeak," Orwell draws precisely this conclusion, showing how, if ideals are simply words, then words like "honor, justice, morality, internationalism, science, democracy" can be banished like so many "false gods," such as the biblical-era replacement of "Baal, Osiris, Moloch, and Ashtaroth" with the Old Testament God.[34]

Habit as such is also endemic to the structures of repetition that generate literary character—we "know" any character, including Orwell's protagonist George Winston, both by his physical attributes, such as the varicose ulcer on his ankle, and his habits, the chief of which is drinking the repulsive gin of the future. The text of *1984* is steeped in the language of habits, both linguistic and otherwise. For instance, it is when he begins seeing his lover Julia that he loses the "habit of drinking gin at all hours," a habit that he drops back into at the end of the novel.[35] Most centrally, in the final exposé of the Party's ideology, the antagonist O'Brien describes the Old World as such as a set of habits, and the Party's endeavor is "breaking down the habits of thought which have survived from before the Revolution," including the "empirical habit of thought" that had led to scientific breakthroughs in other times.[36] The final consequence of this behaviorist doctrine of habit is expressed through dramatic irony: midway through the novel, Winston imagines that the Party's systematic purging and its network of surveillance "could lay bare in the utmost detail everything that you had done or said or thought; but the inner heart, whose workings were mysterious even to yourself, remained impregnable."[37] This distinction between what the Party could find in one's writings and thoughts, on one hand, and one's "impregnable" "inner heart," on the other, is of course collapsed by the novel's ending, in which protagonist Winston Smith is made to betray his lover Julia and to love Big Brother. The "mysterious" workings of the heart are not at all mysterious to the Party, which has mastered the manipulation of habit through various methods of coercive persuasion. The novel's end, in which Smith's "inner heart" is ultimately vanquished, signals the ultimate triumph of a totalitarian regime that has advanced to the point of being built on psychological manipulation and behavior modification. In Orwell's vision, there is no human core—least of all a core called "love"—that lies beyond the purview of propaganda and conditioning.

Orwell, then, presents a narrative that refuses to flinch at the consequences of behaviorist thought, propaganda, and language reform for human beings. The totalitarian technocracy takes advantage of a subject that lacks a core of humanity, a being whose subhumanity lies in its total manipulability. Orwell's dark fantasy combines the methods of propaganda with a radical vision of behaviorism, namely, that everything about human behavior is malleable such that there is no human being beyond behavior. Orwell's *1984* presents a bleak, worldwide obliteration of the "human" subject, in the post-totalitarian state of the future: as such, the text acts as a warning about what is to come if totalitarianism—and the finely tuned behavioral science that is essential to its function—triumphs. Three years after the publication and widespread positive U.S. reception of *1984*, Americans would incorporate Orwell's ideas into an even more fully developed cognitive map of postwar global politics and consciousness.

Brainwashing and the Writers Who Defied It

The "brainwashing" scare, consisting of a handful of popular books and several dozen newspaper and magazine articles in the 1950s, would have a lasting effect on American public discourse, particularly as a mode of describing a kind of mechanical irrationality—unreason within reason—in communism and other dangerous political ideas.[38] Here I trace its beginnings and the scope of its circulation by the early 1960s in the United States, before turning to *The Manchurian Candidate* as an examination of the dialectic of free self and automatic other that the brainwashing discourse helped to establish in Cold War culture. The initial terms of debate about brainwashing circled around the questions of its feasibility and the steps one could take to prevent brainwashing. Following Fromm's lead from the decade prior, these texts meet the terrifying and posthuman new horizons of totalitarian governance and mind-wiping technology from a thoroughly humanist point of view. That is, unlike Orwell's tragic vision of behaviorism's triumph, American propagandists would meet the threats of communism's new technologies precisely by doubling down on the American values that exemplify forms of "human" excess over a behaviorist model of the psyche.

The journalist Edward Hunter coined the term *brainwashing* in English. Although Hunter claims to have heard the term in China, as a translation of *hsi nao*, "wash brain," Timothy Melley has recently discovered that the English term was circulating in American intelligence documents well before Hunter's supposed "discovery" abroad.[39] Hunter had been a foreign correspondent before World War II, and his anti-communist efforts in the Newspaper Guild led to a job with the Office of Strategic Services, the CIA's predecessor, as a "propaganda specialist" for China, India, and Burma during the war.[40] After World War II, he worked as a self-described "roving correspondent" for the Cox newspaper corporation, primarily in China.[41] His first book, *Brain-washing in Red China*,[42] was published in 1951, shortly after reports that UN officers held prisoner in Korea had begun to appear on radio transmissions denouncing America. Hunter's book would receive wider publicity, however, when the threat of brainwashing hit closer to home at the end of the Korean War, when operations Big Switch and Little Switch returned more than seven thousand American POWs to the United States. At that point, Hunter's book would provide the most convenient term for newspaper reporting to describe the possible indoctrination undergone by American POWs in the conflict. Several historians have connected the popularization of brainwashing discourse with the reports of widespread American collaboration in the POW camps, including one *New York Times* article that claimed that as many as one-third of the Americans in captivity collaborated with the enemy.[43] American POWs spoke in radio broadcasts and circulated false reports about American germ warfare, and twenty-one American POWs refused repatriation following the war. David Seed has suggested that the reporters and officials used the brainwashing discourse as a convenient way to silence veterans' dissent at home and abroad: the few soldiers who spoke out against U.S. involvement in Korea could be conveniently deemed "compromised."[44] The brainwashing discourse also arose during the period that Ellen Herman has called the American "Age of Experts," and Hunter and other military and psychiatric professionals used the position of the scientific "expert" to promote a set of polemical notions about Americanness and freedom as part of the Cold War battle for hearts and minds.[45]

Hunter's *Brain-washing in Red China* treats the prospect of mass psychological manipulation as an extension of totalitarian modes of governance. Brainwashing was, in this account, China's most powerful weapon, and at the same time the strongest argument against adopting communism as a system of government in the United States. The book is structured around interviews that detail the reeducation tactics used in newly Maoist China after the 1949 revolution. The interviews describe the party-run schools through which government jobs must be attained,

in which all students were made to keep publicly available daily journals, along with a general environment of suspicion created by secret
police. The interviews themselves detail Mao's extensive program of
thought reform and governmental institutions that would develop, in the
1960s, into the Cultural Revolution. Hunter sets out to show how the
Chinese government could manufacture anti-American sentiment, that
they taught "how hateful America was, and how America was the enemy
of all progressive peoples around the world."[46] Hunter uses the details of
this program to explain the successes of Mao's revolution, and he posits
through these case studies that the Communist Party fundamentally
altered the consciousness of the Chinese people.

Brain-washing in Red China's main claim can be gleaned through
Hunter's comparisons of China with other forms of totalitarian government in Nazi Germany and Soviet Russia. Hunter claims that genocides
and purges in this kind of government will become a thing of the past,
and in speaking of China's "advancement," Hunter describes brainwashing as its highest achievement and claims that only "when [indoctrina-
tion] doesn't work, [would] they resort to the simpler purge system."[47]
The implication in the text is that a successful regime of thought reform
or reeducation must be seen as the culmination of the totalitarian tendency in world politics. Hunter did not invent this continuum between
totalitarianism and thought control from whole cloth, however, as his
repeated mentions of George Orwell's work would attest. The novel *1984*,
too, refers to "totalitarianism" as a phase that precedes the Big Brother
system; the novel's Party operative O'Brien explains, "Even the victim of
the Russian purges could carry rebellion locked up in his skull as he
walked down the passage waiting for the bullet. But we make the brain
perfect before we blow it out."[48] Whereas Orwell wrote about a moment
in which it would be too late to fight the regime of behavior modification, Hunter presents himself as an expert who has arrived just in time
to warn Americans against an evil that, once perfected, will be impossible to fight.

The fullest descriptions of brainwashing itself come in Hunter's two
later texts, from 1956 and 1958, which also mark a shift onto American
soil and subject matter. The 1956 text *Brainwashing: The Story of the Men*

Who Defied It takes the central premise of behaviorism—the notion that most behaviors are not innate and can therefore be conditioned—and draws both political and cultural conclusions from it.[49] Much of the testimony from the "men who defied it" in the books is focused on conditions in prison camps, and on various modes of ill treatment, while the so-called specialist panels are nowhere described in great firsthand detail. In place of this, the text gives a genealogy of "brainwashing," beginning with Pavlov and the behaviorist psychology that "treat[s] the people as animals," describing how conditioned responses differ from innate ones and extrapolating from there.[50] Like Fromm, Hunter will attempt to find the human qualities that allow us to resist being turned into human automatons.

And in his 1958 interview with the House Un-American Activities Commission (HUAC), Hunter describes in more depth the clinical conditions of brainwashing and its potential effects on American culture. There Hunter claims that the behaviorist reform program enabled "the most vicious and sly attacks ever made on the human brain, in which all our great contributions of science and civilization [could be] focused on the upside-down task of making healthy minds sick."[51] Not himself a man of science, Hunter only belatedly discovers that behaviorist conditioning is what he has described in his work, when his fellow professional anticommunist Dr. Leon Freedom, whom he met after the publication of *Brain-washing in Red China,* "clarified the connection between brainwashing and Pavlov's conditioning of dogs."[52] The persuasion and indoctrination techniques he cites include "hunger, fatigue, tenseness, threats, violence, and in more intense cases where the Reds have specialists available on their brainwashing panels, drugs and hypnotism."[53]

Because the ideas of communism are administered in this intentional fashion rather than being willingly adopted, the spread of communism is framed as a pathological affair: it is not a normal kind of idea but one that must be implanted into the populace and spread like a virus.[54] To ensure that communism does not seem like any other idea, Hunter takes up the task of rendering communism unthinkable. With particular clarity in the 1958 HUAC interview, Hunter uses several methods at once to discount communist ideology. It becomes his duty to show why "dialectical materialism" is first and foremost a strategy of "liquidat[ing] our attitudes on what we used to recognize as right and wrong . . . a crackpot theology which teaches . . . that what is right and wrong . . . changes."[55] In addition to depicting communism as an affront to the very concept of morality, Hunter presents Communist Party ideas and activism as a form of bait and switch: Americans can be "seduced into believing something superficial about communism . . . through the semantics of the Newspeak language described by George Orwell," such that they would be made, for instance, to advocate peace in order to weaken America as part of a communist takeover conspiracy.[56]

Likewise, Hunter's 1956 book, in lionizing the soldiers who have resisted brainwashing, acts as much in the way of a prescriptive essay on American patriotic ideals as it does in the way of a warning of what to beware from communists. Hunter takes on the mantle of the psychological expert to make prescriptive claims about who are the best Americans. As Hunter says in the HUAC interview, in the fight against brainwashing, "we have to again go back to characteristics of ours which made us, as individuals, say that what is right is right."[57] Apart from physical and mental toughness on the part of the soldiers, the first main characteristics that fight against brainwashing are "faith and conviction," in many cases religious faith.[58] He tells of religion helping captive POWs over the course of many similar narratives. In fact, Hunter imagines antibrainwashing resilience as something entirely different from mental plasticity or the negative capability that literary scholars prize; in avoiding becoming a dupe to the other side, one must essentially become a dupe to one's own side.

Nowhere is this more clear than in his chapter on "Brainwashing and the Negro," where Hunter's peculiar racial politics again reveal the power of brainwashing discourse to imagine Americanness as a form of humanness distinct from the human automatons of the communist world. Reporting hearsay on the status of African American POWs, Hunter claims, "the Negro had an additional quality" that made him impermeable to brainwashing techniques, "the quality . . . exemplified in Negro songs generally [which shows them to be] without bitterness and without hate."[59] His "heroic resistance to brainwashing" is particularly remarkable because the Negro has no "idea that he had been doing anything special. He had just been himself."[60] The text goes on to report a rather contrived-seeming dialogue, in which the communists attempt to persuade the Negro soldier Roosevelt Lunn that he is a "second class citizen" in the United States and that he should collaborate. Lunn makes replies such as "we're not worrying about the past" and "our position is getting better fast . . . there's a wonderful future ahead for both us and the whites."[61] It is clear from this that the universal appeal of American democracy is essential to the success of the Cold War and of Hunter's brainwashing discourse, and his narrative precludes the inclusion of the "bitter" or dissatisfied American in any way. Perhaps Hunter's most savvy, if still unsuccessful, move is to accuse the Chinese captors of racism against the African American POWs (with a "racist cheese" joke about a "dirty face"), while by contrast, he attempts to show that he understands what is "special" about African Americans and that he knows their love of democracy.

Hunter's paternalistic attitude toward African Americans is clear here, despite his apparent self-image as unimpeachably antiracist. For the history of automatism, it's perhaps less interesting that he fails in attempting to be antiracist than that he makes the attempt at all. This antiracism conforms well to Nikhil Pal Singh's and Kristina Klein's accounts of Cold War racial discourse mentioned in the introduction, wherein Jim Crow seriously undermined the United States's ideological claims to democratic world leadership, particularly after the World War II fight against fascism and the opening of the Nazi concentration camps. Hunter's propaganda itself is not explicitly racist, as it describes an enemy that cannot at any rate be easily represented through the nationalist and racial propaganda that had proven useful in the early twentieth century. Where the racialized body reemerges instead is in the deep essentialism that Hunter uses to imagine African American resistance to brainwashing. Where Orwell had not been inclined to imagine a core of personality that propaganda and torture cannot touch, it is precisely this core of personality on which Hunter's plan for resisting brainwashing propaganda relies. The biologically or culturally essential qualities that Hunter here attributes to African Americans work in parallel to the faith, patriotism, and steadfastness that he values as the human essence in other American soldiers, as aspects of being American—if one is indeed American *enough*—that the brainwashers cannot touch.

Samuel Fuller's 1963 film *Shock Corridor* shrewdly reverses Hunter's line of argument on this count by placing the burden of susceptibility to brainwashing on the soldier's native education and environment. *Shock Corridor* features a brainwashing victim and veteran, Stuart, who has been placed in a mental asylum for reenacting a Civil War fantasy. Stuart had been easily converted to communism as a POW because he had grown up eating "bigotry for breakfast and ignorance for dinner," with "no knowledge of [his] country," the implication being that his lack of the sufficient patriotism had been no accident. His ignorance and lack of patriotism do not reflect on Stuart so much as they put forth a criticism of the country that had done nothing to make him "proud of where [he] was born." Although this character is eventually converted back from communism (as a plot device to bring him into the eponymous asylum ward), he acts out a psychotic fantasy wherein he is forced to reenact the Civil War. That fantasy sequence, incidentally reprised in the second season of David Lynch's *Twin Peaks* (1990–91), serves as a reminder in Fuller's film of the injustices in a country to which Hunter would encourage unflinching and even unthinking "faith and conviction." And indeed, in his report about the African American soldier, Hunter seems to realize that unflinching faith and conviction in America may also depend on forgetting the domestic race issue, hence Roosevelt Lunn's odd-sounding statements cited earlier that he is "not worrying about the past" and that "our position is getting better fast."

Throughout Hunter's writing, the notion of brainwashing assumes the essential sameness—through the term *totalitarianism*—of fascism and communism.[62] The most cogent definition of totalitarianism, as a category into which fascist and communist movements fall, comes from the 1958 second edition of Hannah Arendt's *The Origins of Totalitarianism*.[63] There Arendt states the equivalence between Nazism and communism as the act of forcing a "law of Nature" (Darwinism) or a "law of History" (Marxism) into being, an act that "claim[s] to transform the human species into an active carrier of a law to which human beings otherwise would only passively and reluctantly be subjected."[64] In other words, Arendt posits that the leaders of these countries attempted, through their subjects, to expedite evolution itself or, in the case of the communists, a Marxian vision of history. The supposed Chinese focus on indoctrination forces this Marxist law of history onto its subjects through psychological manipulation, what Hunter will call a "weapon for conquest intact of peoples and cities."[65] This threat of psychological destruction *in the place of* physical destruction clearly changes the terrain on which the United States would want, in Hunter's view, to fight the Cold War.

Over the course of the 1950s and the early 1960s, the conceptual problem of "brainwashing" grew legs, and I will trace some of its destinations before turning to *The Manchurian Candidate*. Of the many brainwashing experts of the 1950s, the psychiatrist Joost Meerloo offered the most explicitly Orientalist account of the psychology of unfreedom. He posits that there is in the East a "cultural predilection for totalitarianism," and he variously cites "Judaeo-Christian ethics" and the "ethics of our own Western civilization" as the sources of American "individualism" and Americans' tendency to "evaluate situations primarily in terms of [their] own consciences."[66] The "Oriental" individual is, conversely, "not a separate, independent entity" in the same way, owing to her religion and customs.[67] Meerloo elsewhere echoes the ideas of Hunter, Fromm, and Orwell in his warnings about the "robotization of man," "*automatic thinking* that is tied to [special] words," and the "semantic fog," "logomania," and "logocide" that occur in the totalitarian state.[68] Ideas like these spread alongside *1984*'s long best-selling run in the United States, and most of the dozens of newspaper and magazine articles about brainwashing featured some combination of them.

The word *brainwashing* appears nowhere in Vance Packard's epochal *Hidden Persuaders*,[69] on the power of subliminal advertising, perhaps because Packard wanted to differentiate himself from the often wildly speculative writings on communist mind control techniques. Nevertheless, others were quick to take up the connection between media technologies and the inducement of totalitarian psychology. Marshall McLuhan, for instance, took up the problem of brainwashing as part of his broad conception of "media" in *Understanding Media*.[70] In a story he draws from Philip Deane's *I Was a Captive in Korea*,[71] McLuhan discusses the POW's success in defying brainwashing through games and stories: Deane "could feel his "thinking processes getting tangled, [his] critical faculties getting blunted . . . then [the communists] made a mistake. They gave us Robert Louis Stevenson's *Treasure Island* in English. . . . I could read Marx again, and question myself honestly without fear. Robert Louis Stevenson made us lighthearted, so we started dancing lessons."[72] As a metaphorical "extension" of the human, McLuhan asserts, the game provides both "models of our psychological lives" and a way to "release tension."[73] This admittedly strange parable of McLuhan's emphasizes how connections to the outside world are in fact a central part of an individual's psychological makeup, such that even books make visible a world outside the cell. Following this logic, the prisoner in solitary confinement is no longer quite the same man he was before. McLuhan offers, then, a slightly different solution from Hunter's "faith and conviction," but it's striking that even in 1964, well after the height of the brainwashing scare, even as respectable a thinker as McLuhan considers brainwashing an interesting problem. He even seems to take some measure of pride in his method of what one article from the 1950s *Saturday Evening Post* had called "baffling the brainwashers."[74]

And the conceptual problems associated with brainwashing were taken seriously in political theory as well. The British philosopher Isaiah Berlin considers the problem of the ethics of "brain-washing" in his major 1958 lecture "Two Concepts of Liberty."[75] As Berlin remarks, any kind of brainwashing scheme would be one of "shaping [citizens] against their will to your own pattern [and moreover] all thought-control and conditioning, is, therefore, a denial of that in men which makes them men and their values ultimate."[76] In this tangent to the lecture's main concepts of positive and negative liberty, Berlin's point is that the technocratic management of others violates a principal tenet of Kantian ethics. If one uses one's subjects "as means for my, not their own, independently conceived ends, even if it is for their own benefit," it is, "in effect, to treat them as subhuman, to behave as if their ends are less ultimate and sacred than my own."[77] Berlin goes on to discuss in a footnote how the communist intellectual Nikolai Bukharin betrays this tendency of thought in his own writing. Bukharin had written in the 1920s about how the Communist Party should "mould . . . communist humanity out of the human material of the capitalist period," and it is the notion of "human material" that sticks out to Berlin as an ethical problem. (It likewise meshes well with Arendt's contemporaneous definition of totalitarianism.) And for Berlin, the problem resurfaces as "brain-washing," which he sees as the highest expression of that technocratic and instrumentalizing approach. In going to the trouble to make the rather obvious point that brainwashing is a bad thing to do to people, Berlin draws attention to the matter of how the totalitarian technocrat tends to *see* the individual subject: not as an ethical actor or as an end in himself but as "human material."[78]

That ethical doubleness of the subject, as end-in-itself or as programmable raw material, was also the main conceptual problem that the literary and cinematic figure of the human automaton addressed in the mid-century period. Robert Heinlein's novel *The Puppet Masters*[79] considered this problem of human material extensively, and it was also among the first and most popular fictional treatments of a massive brainwashing campaign avant la lettre, on U.S. soil and perpetrated by aliens. The takeover begins at a UFO crash in Iowa, and the secret agent protagonists fight the rapid spread of the Titans, parasites that latch onto victims' upper backs. Once the secret agents reclaim the White House from the "slugs," they have to implement "Schedule Bare Back" as part of "Operation Parasite," whereby everyone has to keep her back in plain view.[80] The fiction is then structured around the Titans' ruses—from prudish protests to the shirt law to strategically timed transmission breaks in videoconferences with the government—that keep the protagonists guessing about who at any given time is "hag-ridden."[81] As Leerom Medovoi and others have noted, *The Puppet Masters* anticipates the brainwashing discourse by a couple of years, particularly in its discussions of Russia, where "Stalinism seemed tailormade" for the Titans, because the "people behind the Curtain had had their minds enslaved and parasites riding them for three generations."[82] But the bare fact of this allegory aside, the fiction rehearses many of the philosophical and practical questions that Hunter, Berlin, and Meerloo would formulate later in the decade. How can we be sure that we know someone has or has not been brainwashed? What character traits might make us more or less susceptible to brainwashing? And when we find subjects we think have been brainwashed, do we not have a duty to treat all humans as ends in themselves, even and especially if they are being used as "human material" by others? That choice of how to see the subject of brainwashing would also be at the center of the films that featured brainwashing, including John Frankenheimer's *The Manchurian Candidate*.

The Veteran on the Screen

For the segment of Hollywood producing war films, the Korean War presented no small challenge in representing the enemy, and many of the films were commercial flops. Writing about all eighty-eight Hollywood films made about the Korean War from the 1950s to the 1990s, Paul Edwards notes the apparent difficulties: that the recently Allied Chinese seemed, in the first place, like the "wrong enemy" and that many Korean War films were unsuccessful because they were still working within a World War II framework, as a "fight between nation-states."[83] "Communism," Edwards writes, "was much more easily represented as the mind-bending ideology whose crimes were reflected, later, in a whole series of 'brainwashing' films."[84] The visual style of the brainwashing film would be borrowed from the machine-age automatons and early totalitarian automatons discussed in the Introduction, which seemed ready-made for the depiction of both programmed bodies and brainwashed minds.

The Manchurian Candidate has come to be acknowledged as one of the great films of the Cold War, despite its having dropped out of circulation following the assassination of John F. Kennedy.[85] Many readings of the film have tried to place the politics of the film in relation to the politics of McCarthyism and gender in the Cold War.[86] That is, they tend primarily to resolve the tension between the film's "surface" politics of anti-anticommunist satire and the more conservative undercurrents of the film's plot structure and gender politics. The most convincing of these readings end with the figure of Raymond Shaw's unnamed mother, whom they rightly name as the film's scapegoat. Onto Shaw's mother are heaped all the negatives of McCarthyism, communism, and the over-bearing mother of the Cold War nuclear family, before she is killed off without an adequate resolution or explanation. Though I certainly accept the terms and conclusions of this interpretation, these readings of the film neglect the film's most important narrative motor—the mechanism of brainwashing and the display of Shaw as a human automaton—and, consequently, overlook much of the intellectual work the film performs. Instead, with the work of Fromm, Hunter, Berlin, and Arendt in mind, we can see that *The Manchurian Candidate* grapples with the same ethical and practical dimensions of brainwashing at the level of cinematic form, through its influential use of the human automaton figure.

The heightened confusion at the end of *The Manchurian Candidate* shows insufficient motivation not only for Shaw's mother's politics but also for the brainwashing plot itself. That is, pace Richard Condon, author of the novel on which the film is based, it is wholly implausible that the Chinese or anybody would have made a plan like this one: the communists want to assassinate a presidential candidate so that the vice presidential candidate (John Iselin, a parody of Joseph McCarthy) can take over and *almost* certainly be elected president.[87] The projected President Iselin will be under the influence of his badgering wife, Shaw's mother, who is a high-ranking communist official, and communism will spread to the United States. Why would the communists need to employ or develop a brainwashing technology to have this ruling couple's son, Raymond Shaw, be the assassin? And why—another crucial point for much of the film—does the assassin need to have won a Congressional Medal of Honor to carry out this plot?[88] There are two interrelated answers to these questions: in the first, to take a cue from the plot's similarities to *Hamlet*, the film's plot is primarily concerned with the psychology of its protagonists, such that the political details are of lesser importance. The psychological problem in question is what the film's vocabulary calls "combat stress" and what has since been called PTSD.[89] The second answer to the question of the plot's problem, and the most directly intuitive answer, is that the plot is itself an excuse for putting brainwashing on gratuitous display. The thoroughness and complexity of this display place *The Manchurian Candidate* among the richest cinematic uses of the human automaton. As with much of the genre, the brainwashing victims, Shaw and Marco, must test both the limitations of the mechanism of brainwashing and the limit cases of what we consider to be human.

Reading veteran psychology as the structuring element to the narrative lends the fullest and simplest explanation of its events and sequencing. The narrative plays out as two parallel stories, one of a cracked-up veteran who works through his problems and another of an outwardly sane veteran who unpredictably blows up. The first of these is Captain Bennett Marco, played by Frank Sinatra, and his piece of the narrative is recounted as the working-through of a symptom. Since his tour in Korea, he has been tormented by a vivid, recurring nightmare about his capture there. In this dream, he sits passively by as Shaw strangles their fellow soldier Ed Mavoley, as part of a demonstration of Shaw's successful brainwashing. In the same dream, master brainwasher Yen Lo orders Marco to recommend Shaw for the Congressional Medal of Honor, citing a phony story of Shaw's heroic acts in a fictitious battle. In the present of the film's beginning, Marco has done so, and he dutifully recites the set phrase about Shaw—that he is the "kindest, warmest, bravest, most wonderful human being [he has] ever known in [his] life"—despite not having liked him personally. All is not right, and Marco's nightmare bleeds into his waking life. He becomes unable to work effectively as an army officer, and he develops nervous tics and something of a sweating problem. As the film progresses, these become a full-blown neurosis: in a memorable train scene, he is pathologically unable to light a cigarette, dropping the match each time it approaches the cigarette. The persistence of the nightmare scene, and of the unsolved mystery behind it, stands between Marco and any rest or satisfaction, even that of smoking. In fact, the recurring dream explains the cigarette problem to the viewer: Yen Lo's men, in the brainwashing display scene, had filled the soldiers' cigarettes with "yak dung," and the brainwashed men smoked them happily, unable to tell the difference. As the memory surfaces, apparently, some part of Marco protests against the cigarette, against both the false satisfaction of the sham smokes and the possibility of repeating the awful scene while fully conscious.

The mystery at the bottom of this representation is *how* Marco resists the brainwashing. In Hunter's texts, the resistance comes from either an excess of patriotism or a racial essence; Marco could certainly be said, as a hip, all-American Italian American, to fit within this schema, particularly if we consider him as one of Norman Mailer's "white negroes." The film's own terminology and self-consciously Freudian vocabulary supplement this account of a natural resistance, as the resistance to brainwashing acts as the return of the repressed. Either way, this inner conflict plays out in perfect accord with the current *Diagnostic and Statistical Manual of Mental Disorders* listing of the symptoms of PTSD: recurring nightmares, nervous tics, and the victim's inability to do anything closely associated with the event.[90] Marco's superiors in the army do not believe the story about his brainwashing dream, and he eventually breaks down, gets fired, and finally finds himself under arrest following a karate-style fight with Shaw's Korean houseboy. He is only cured when the African American soldier from his platoon, Al Melvin, corroborates the story—the two soldiers miraculously identify the figures from their dreams in separate photo lineups with army intelligence. While this coincidence confirms the truth of Marco's experiences, it shows badly for the rest of his platoon—in both the novel and film, it is the two non-white (or not-quite-white) soldiers who mysteriously resist the brainwashing, where the white soldiers are all successfully brainwashed.[91] As in Hunter's *Brainwashing and the Men Who Defied It,* race reappears as a bodily essence that exceeds brainwashing's ability to take over the mind. The triumph of Marco's photo matchup with Melvin is, of course, the moment where Marco's PTSD is cured, and his nightmares presumably cease. Or rather, as in any detective story, Major Marco overcomes the effects of the crime by asserting rational control over them, by breaking the code. The emblematic and climactic moment of this code breaking (used in the film's publicity posters) is his discovery of the trigger for the brainwashed assassin Shaw's activation mechanism, a card bearing the queen of diamonds. After Marco's superiors are convinced of his story, his nightmares and other symptoms subside, and he smokes a very satisfying cigarette to celebrate.

Raymond Shaw, the brainwashed assassin, plays out the opposite scenario for the viewer. Over the course of the film, he is the cold, unlovable, and supposedly successful veteran (as a Medal of Honor winner and assistant to a famous journalist) who goes on a homicidal rampage, killing, in cold blood, his boss, his father-in-law, his new wife, his stepfather, his mother, and himself. From this list, it seems quite clear that Shaw is a bad veteran, a broken man of the type who haunts the pages of Michael Herr's *Dispatches*[92] a decade and a half later. Nevertheless, the film goes to great lengths to assign different motivations to each of his killings: first, a test of his continued efficacy as a brainwashed assassin (his editor boss); second, a political maneuver (father-in-law, Senator Jordan); third, the application of the "no witnesses" clause of his orders (wife, Josie); fourth, his breaking free and stopping the political plot (his mother and stepfather); fifth, apparently out of despair, suicide. We might also add to this count the two murders from the brainwashing display in Manchuria, where he kills the two fellow soldiers he "dislikes the least," at the behest of the "Russian brass," who are anxious to see him kill. The film thereby dooms Raymond to the tragedy of his own death and that of everyone close to him (excepting Marco, the witness) yet deprives the tragedy of its common denominator, Raymond's own will. This heretofore neglected current within Marco's and Shaw's subplots in the film strongly evokes the problem of the reintegration of the veteran with PTSD—"combat stress," as the film calls it—and we might say the film resolves this problem by eliding it with the fundamental problem motivating the Korean War itself: the spread of communist ideology. The film thus tracks its protagonists' value as Americans through their pursuit of a cure; accordingly, the subplots of Marco's and Shaw's romances show their rhymed sweethearts (Rosie and Jocie) as means to the ends of their recoveries.

While this thread of combat stress and recovery helps to explain the film's elaborate plot, it also provides the alibi for the film's repeated and often brilliant scenes of brainwashing. The brainwashing genre, as David Seed has pointed out, is defined by a flaw in the brainwashing mechanism such that the plot is centered on the struggle of the brainwashing victim against the total takeover of his own mind.[93] At the level of narrative, at least one victim of brainwashing has to be able to fight the influence of brainwashing, or there is no way to build tension or sympathetic identification with the character. Much of the film's captivating quality—its sharp, at times bizarre, dialogue aside—can be explained by its display of brainwashing, which is expressed as the toggling between conscious decision-making ability and a loss of that ability. This toggling is also an alternation between a subject wholly determined by behaviorist conditioning and a subject that exceeds the boundaries of that conditioning. The display of brainwashing, as it depends on time, is essentially responsible for all the visual suspense in the film. The very premise of the action, then, allows us to judge the film's two protagonists by their threshold for the morally abhorrent: we wait and hope that each character has a kernel of American free will, which will have the strength to erupt in violent protest of the atrocities being committed. The first brainwashing scene sets this expectation for those that follow it. The climax of this sequence, which features some powerful editing, starts in Marco's dream, where Shaw is strangling Ed Mavoley, then cuts to Marco, brainwashed, lackadaisically yawning. Then, just before Mavoley dies, Marco wakes up from the nightmare, screaming and sweating. Rather bizarrely, it seems that the brainwashed Marco is catching a contagious yawn from the gasping soldier. This doubly inappropriate response to the death of one of his men prompts the dreaming Marco to wake up in protest. The kernel of free will triumphs, and Marco remains our sympathetic protagonist. The same aspect of Marco that had known "deep down" that he hated Shaw, and that began to repeat the dream in the first place (he is supposed to have forgotten it), has awoken, screaming in horror at the monster he has helped to create. It is here that the visual language of the film first communicates the contrast between the brainwashable subject as a human automaton and the individual as an ethically accountable being. In the "brainwashed" Marco, there is a flat, affectless tone—he answers questions matter-of-factly, as though he is completely detached from the situation and devoid of ethical investment in the situation he faces. The concept of brainwashing, and the idea that it might be implemented, relies on the behaviorist assumption that the subject is a set of manipulable behaviors and nothing more. This scene's strategy for

Frank Sinatra as Bennett Marco and Joe Grey as Ed Mavoley in John Frankenheimer's *The Manchurian Candidate* (1962). Raymond Shaw (Lawrence Harvey) strangles Ed Mavoley in a demonstration of Dr. Yen Lo's brainwashing technique, as seen in Marco's recurring nightmare. In the dream, the brainwashed Marco yawns at the gruesome scene, but he wakes up from the nightmare screaming.

representing freedom, the possibility of ethical action, is Marco's depar-
ture from the predictable program of the dream. Marco's resistance in
his nightmare, the flaw in the mechanism's design or implementation,
comes to stand, in the film, for the point of excess that makes him an
ethical subject as well.[94]

The most important cinematic aspect of the "sleeper agent" device in
the film is that it allows for the insistent repetition of the transition from
consciousness to automatism and back again. Two subjects alternate:
the humane if largely Freudian subject of ethical accountability and the
manipulable subject of a hypothetically perfected behaviorist psychology. Raymond Shaw's subplot, in formal terms, consists of a long series
of tests and one "misfire" of the brainwashing mechanism, and the narrative seems to want to disprove the myth, along with the Pavlov Institute's Yen Lo, "that no brainwashed subject can be made to commit acts
that are morally abhorrent to him." Whereas Dr. Yen and, in the end,
Shaw himself are both conveniently interrupted when they want to tell
about how the brainwashing works (some combination of "chemicals
and lights"), various characters insist, seemingly at every turn in the
film, that the mechanism be either used or demonstrated. In addition
to assigning different motives to each killing, the discourse of the film
insists that the mechanism be shown again and again. The Soviet general requires that Shaw kill at Yen Lo's display scene, only to insist that
he kill another man moments later. Yen Lo later argues with Vilkov in
favor of another test of the mechanism, who complains that Shaw must
be tested, because he "has not killed for over *two years*"—which is, for the
viewer, about twelve minutes. The targets grow systematically more emo-
tionally close to Shaw, ending, before the film's climax, with his new
wife, Jocie, whom he kills under the "no witnesses" clause of his orders.
As a principal structuring element of the narrative, the brainwashing
display prompts the viewer to ask each time if he'll really go through
with it, each time judging Shaw against the dreaming Marco's inability
to stand by as a killing occurs in the demonstration scene.

The final test of the mechanism coincides with the climax of the film,
in which Shaw is supposed to shoot the presidential candidate. Captain

Marco has, in the scene just beforehand, "gotten to" Shaw by showing him the trigger card from an all-queen-of-diamonds deck and making an all-American incantation: "we're bustin' up the joint, tearin' out all the wires. . . . If someone asks you to play a game of solitaire, you tell 'em, 'Sorry, buster, the ballgame's over.'" These words are intended to reverse-program Shaw, even as the remark about "wires" would suggest that he has to approach Shaw as a machine, telling him, "You don't work anymore" rather than positively reaffirming his humanity. As Shaw goes into the booth to perform the assassination, it becomes abundantly clear that there had been no secure criterion for telling Shaw's brainwashed state from his normal state. The viewer then goes on to hang in suspense, wondering through this gun-sight scene whether the brainwashing has been undone. He sets his sights on the candidate the communists have told him to shoot, while we are forced to wait for a key sentence in the speech. These camera angles give a tense spectatorial identification to the scene. The viewer is confronted with the inaccessibility of the other mind's consciousness during a series of crosscuts between a medium shot of Raymond and Raymond's own view through the gun sights. The tension is relieved when he shoots the villains, Senator Iselin and his own mother, and then himself. It is only then that the audience becomes certain that Raymond has been cured, and his suicide forecloses the problems of his legal responsibility or continued life. Here, as with the images of Major Marco waking up yelling and the image of cracking the code, the cure is eminently representable: it is the strongest moment of spectatorial identification, embedded in a release of tension in the narrative. The traumatic loss of the freedom of rational choice, conversely, removes the brainwashed subject from our democratic community of free subjects, and we see that Shaw could not exercise the free choice necessary for democratic statehood.

One in every dozen or so shots in *The Manchurian Candidate* features a portrait or bust of Abraham Lincoln, which suggests, perhaps, an emblem of what stable or traditional notions of freedom, and American-ness, might have once meant.[95] Relative to this historical marker of emancipation, the film documents the shifting definitions of democratic

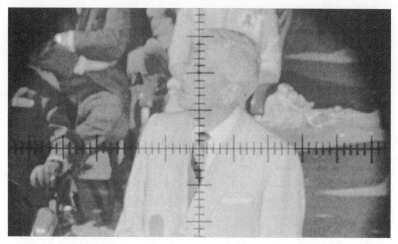

As Raymond Shaw (Lawrence Harvey) is about to commit the final assassination in John Frankenheimer's *The Manchurian Candidate*, viewers hang in suspense as they look down the gun sights along with Shaw, unable to tell whether his brainwashing mechanism is still operative.

community and freedom in the Cold War period. The notion of brainwashing as it was used in Cold War propaganda allowed Americans to imagine that the citizens of totalitarian states had been somehow conditioned, such that they became masses of human automatons, will-less and therefore less than human. The film's work as tragedy (as a revision of *Oedipus,* the *Oresteia,* and *Hamlet,* albeit with a flat, unfeeling protagonist) is accordingly to banish the brainwashable subject from America. The American who gets to remain, in the ethical calculus of the film, is the fully human, free, and morally accountable Marco. The film's politics come down, at the level of form, on the side of Hannah Arendt and Joost Meerloo, in defining the psychology of American freedom over and against totalitarian unfreedom.

That dialectic of freedom and unfreedom becomes slightly less stable, however, in the final instance of the film's elegy scene, which poses *The Manchurian Candidate* as a veteran "buddy" film. In this sense, Marco's faithfulness to Shaw seems exemplary; motivated by a search for the truth of the situation, he acknowledges Shaw as a fully human being despite the brainwashers' attempt to reduce him to a biological set of programmable reactions. Marco imagines that his compromised comrade may, for reasons unknown, be redeemable. Had he really wanted to avert tragedy, he would have simply killed or detained Shaw at the film's end. Instead—and this is, of course, also an imperative of the action film genre—he must save him. The film ends with Marco's elegy for Shaw, a meditation on Shaw's Congressional Medal of Honor. The medal, which had not played a role in the film's central assassination plot, fulfills its ultimate function when it embodies Marco's grief, who creates a new version of Shaw's Medal of Honor citation: "made to commit acts too unspeakable to be cited here, by an enemy who had captured his mind and his soul. He freed himself at last, and, in the end, heroically and unhesitatingly gave his life to safe his country. Raymond Shaw." The connection between the two men makes for an unlikely buddy film, considering that all the men "hated" Shaw. Marco's quest to save Shaw bespeaks the same kind of commitment to humanity as Isaiah Berlin's resistance to treating others as "human material" had done. The paradigm that allows Edward Hunter and other writers in the 1950s to speak of the communist as wholly conditioned or "brainwashed" allows one to speak of the ideo-logical other as completely foreclosed from ethical acknowledgment, as forms of "human material" not so unlike that of the totalitarian state. *The Manchurian Candidate*, in its sympathy for Shaw, foregrounds the possibility, and the difficulty, of acknowledging the other's dignity when neither freedom nor dignity can be seen.

CHAPTER TWO

Anti-institutional Automatons

New Left Reappropriations of Automatism

Before arriving in Nurse Ratched's mental ward, Randle Patrick McMurphy had received a "Distinguished Service Cross in Korea, for leading an escape from a Communist prison camp."[1] Ken Kesey's protagonist in *One Flew over the Cuckoo's Nest* follows in the footsteps of *The Manchurian Candidate*'s Marco and Shaw, as a Korean War veteran trying to resume his civilian life in the United States. A while after his return from overseas, McMurphy manages to parlay a prison sentence into what he thinks will be a lighter sentence in the mental ward. The implication of the POW backstory here is that the American authorities, embodied in Nurse Ratched, constitute the more formidable foe. Midway through the novel, Nurse Ratched orders electroconvulsive therapy (ECT) for McMurphy in an unsuccessful attempt to alter McMurphy's personality and to exert complete disciplinary control over his rebellious attitude. Milos Forman's 1975 film version uses this shock treatment for a scene of intensive struggle between the disciplinary apparatus and the individual, in the now-iconic image of McMurphy, played by Jack Nicholson, stuttering and shaking during the procedure. Minutes later, the film plays the ECT's aftermath for a scene of comedic automatism, where McMurphy stumbles in to see his wardmates, pretending to be a vegetable before bursting into "gotcha" laughter. The other inmates join in his laughter as McMurphy's enactment of Bergsonian mechanism returns to his usual performance of freedom and defiance within the ward. The institution has attempted to condition, discipline, or suppress, but some unfathomable excess of McMurphy's individuality—like Marco's in *The Manchurian Candidate*—exceeds the institution's power to subdue him. When it comes to lobotomies, however, the novel's narrator warns the reader that the institution's measure of success diverges more dramatically from that of the inmates: a calmer patient postoperation might be "a success, they say, but I say he's just another robot for the Combine and might be better off as a failure like [botched lobotomy patient] Ruckly sitting there fumbling and drooling over his picture."[2] Throughout the text, robots and machines play a prominent role in Bromden's hallucinations of America's social machinery, the Combine, against which McMurphy stands as a model of disobedience, trickery, laughter, and humanness. When McMurphy, having incited a rebellion, receives his own botched lobotomy, his comrades are unable to recognize the shell of their former idol. Bromden exclaims, "There's nothin' in the face. Just like one of those store dummies," and the novel devotes most of a page to the slow dawning of recognition among his comrades, playing it for uncanny effect.[3] McMurphy's individuality has been extinguished, setting up a tragic ending for this now-subhuman subject of the institution.

As I argued in chapter 1, the cinematic and literary figure of the human automaton—Frank Capra's marching denizens of the "Slave World," Orwell's "quack-speakers" of totalitarian language, Edward Hunter's and Joost Meerloo's "brainwashed" POWs in the Korean War—had provided a visual and textual paradigm for representing mental unfreedom in the 1940s and 1950s, one fit for portraying the new enemy of "totalitarianism" during and after World War II. During those decades, writers across a number of disciplines and professions worried that aspects of U.S. society made it susceptible to totalitarian conformity, including Edward Hunter, William Whyte, David Riesman, Hannah Arendt, Theodor Adorno, and Erich Fromm. But by the 1960s, the same rhetoric of anticonformist individualism, and the same image of totalitarian automatons, was often being mobilized against the U.S. establishment itself. The career of this image highlights a continuity between the antitotalitarian culture of the 1950s and a significant strain of the progressive culture of the 1960s, a continuity implicit in McMurphy's transition from POW

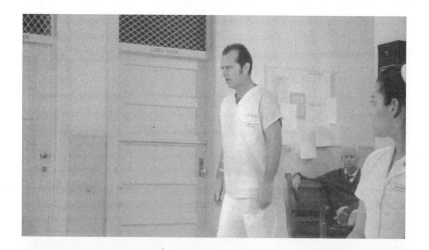

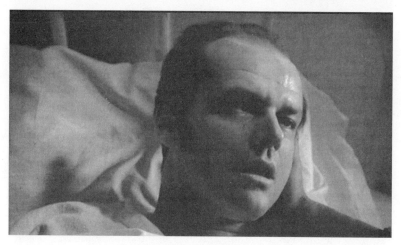

Jack Nicholson as Randle P. McMurphy in Milos Forman's *One Flew over the Cuckoo's Nest* (1975). McMurphy fools his fellow mental asylum inmates into thinking his electroconvulsive therapy turned him into a zombie, and they all laugh in the following frame. Later, he is shown after a lobotomy that has rendered him speechless and expressionless, and his head bobs lifeless in his friend's hand.

anticommunist to antipsychiatry rebel. That continuity suggests a different narrative about Cold War culture from Alan Nadel's metaphor of 1950s conservatism as a "containment culture" that eventually gives way to the watershed of 1960s progressivism.[4] In fact, a wide variety of texts from both decades contested the meanings of freedom and unfreedom, all using the same narrative and rhetorical strategies that surround the human automaton as a figure. An anti-institutional *ethos,* often most concisely expressed through automaton imagery, enables a significant strain of progressive rhetoric in the 1960s, and one that has afterlives, too, in the libertarianism and commercial culture of the 1970s and 1980s.

If *One Flew over the Cuckoo's Nest,* as Mark McGurl has observed, was written while Kesey was a mental ward night watchman and a creative writing graduate student, a novel "produced in the shuttling from one kind of institutional space to another," it is also a novel that is sustained by the comparisons between institutions on different scales and within different forms of social system.[5] The key difference between the ward and the writing workshop implied in McGurl's analysis is that the workshop is an "open system"—sustained by interchanges with the outside world—whereas the mental ward depends for its proper function (and metaphorical significance) on the illusion of total closure, where no interlopers beyond the inmates, doctors, and guards can be allowed.[6] The institutional microcontrol exercised by the mental ward in *One Flew over the Cuckoo's Nest* is mirrored in the global reach of the "Combine" that Kesey's narrator posits as a corporate–industrial–institutional complex of society at large. "The ward," Chief Bromden tells us, "is a factory for the Combine," and the authoritarian Nurse "dreams of . . . a world of precision, efficiency, and tidiness like a pocket watch with a glass back."[7] Escaping the mental institution and escaping the Combine are two different matters altogether, but the former can stand in for the latter easily enough in a narrative of escape. *One Flew over the Cuckoo's Nest* and many texts like it draw sustenance from the fantasy of escaping such a "Combine" of the 1950s and 1960s. For instance, a similar institutional homology could be said to structure the three main sections of Allen Ginsberg's "Howl," where *Metropolis*'s "Moloch" stands in for industrial modernity's institutions writ large, the "best minds . . . destroyed" for failed lines of flight and freedom, and the metaphorical state of being always "with you in Rockland"—that is, an asylum inmate—as the inscription of the small institution onto society as a whole.[8] Ralph Ellison's *Invisible Man* would prove influential in this regard by featuring a "factory-hospital," whose combined-institutional facets would be mirrored in its depiction of the Communist Party as a space of "discipline," "science," and "machinery."[9]

While critics tend to be most familiar with Michel Foucault's analyses of the isomorphism between social institutions—"factories, schools, barracks [and] prisons" all resemble both one another and society as a whole—these literary texts anticipate this aspect of *Discipline and Punish* in an instructive way.[10] The texts I examine in this chapter are thus proto-Foucauldian insofar as they imagine the form of institutional discipline as something that insidiously interrupts the everyday spontaneity and freedom of ordinary social interaction. In a fine-grained analysis of the lived experience of institutions, the sociologist Erving Goffman influentially described in 1961 what he called the "total institution," in ways that also resonate with these novels and films from the 1950s and early 1960s. One of the key features he found in total institutions—such as mental asylums, prisons, schools, monasteries, and other social spaces that effectively enclose their populations—is that they rob their inmates of a feeling of autonomy. They consistently "disrupt or defile precisely those actions that in civil society have the role of attesting to the actor and those in his presence that he has some command over his world— that he is a person with 'adult' self-determination, autonomy, and freedom of action."[11] *One Flew over the Cuckoo's Nest* rehearses many of these deprivations of freedom and means of infantilization, though the most emblematic scenes of unfreedom exceed the realism of Goffman's inquiry to make use of the aesthetic power of automatism. Those scenes of ECT and of mindless motion most effectively and memorably stage McMurphy's transformation into a subhuman object.

The total institution and the image of the human automaton stand as complementary concepts in postwar U.S. cultural texts, a set of narrative conventions that imagine the full success of discipline, even sublime discipline, through its total control over the subject. In this context, then,

the figure of the human automaton stands in for this disciplinary sublime, as the emptied-out behaviorist subject of the total institution, as
with the tragically lobotomized McMurphy. This chapter describes some
of the ways in which automatism enabled a progressive rhetoric of individualism in the decades following World War II. One of the earliest and
most sophisticated sets of anti-institutional automatons populates Ellison's *Invisible Man*, probably the most sustained and influential use of
both automaton imagery and ECT in postwar U.S. literature. In Ellison's
hands, the scene of automatism was usefully reversible: instead of depicting the body of an inscrutable enemy as that of an automaton, Ellison
proposed his own protagonist as the automaton. By showcasing the reductive gaze of the postwar institution upon African Americans, Ellison
sought to expose a pattern of instrumental treatments of the individual.
For Ellison, the automatism of his protagonist went hand in hand with
his invisibility under the institution's gaze, and *Invisible Man* provides
a particularly sustained ethical investigation of automatism and institutions. A decade later, Kesey would borrow Ellison's image of ECT, and
the following year, Betty Friedan mobilized images of totalitarian automatism in an iconic work of activist nonfiction, *The Feminine Mystique*.[12]
Friedan for me marks the apotheosis of this rhetoric in that she successfully depicts the housewife's condition as one of dehumanization by
strategically framing the household and the marketplace as metaphorical total (and sometimes totalitarian) institutions. Charting the spread of
this rhetoric into Silicon Valley cool, this chapter also takes up the commercialization of this countercultural institutional automaton imagery,
through the Super Bowl advertisement for Apple's Macintosh computer
in 1984. Across the wide range of rhetorical uses of automatism in antipsychiatry, antiracism, feminism, and commercial culture, I hope to suggest a trajectory from Ellison's serious, modernist use of the automaton
to its camp afterlives in the 1970s and 1980s. By tracing some of these
anti-institutional automatons through U.S. culture, I hope to contribute
to a project of demystifying and historicizing the institution as a cultural
idea and to explain the appeal of imagining all institutions as panop-
ticons, as spaces of coercion and unfreedom, and as locations inside of
which individual agency seems all but impossible. As such, this chapter

aims to draw attention to the role aesthetics might play in the political and academic abhorrence for the institution as a form, which bleeds into contemporary rhetoric that addresses the welfare state, unions, neo-Keynesian economics, single-payer medicine, and the public university.[13] Even as I trace the influential currents of progressive rhetoric in these uses of the automaton figure, they also suggest, in a thread to which I return in the conclusion to *Human Programming*, the need to imagine individual and collective agency without resorting to the posture, and the allegory, of the individual rebelling against the institution.

Ralph Ellison's Automatons

Ellison was particularly resourceful—and influential—in adapting the automaton imagery of antitotalitarian and "mass man" rhetoric toward his own political ends. The walking iron bridge with which *Invisible Man*'s main plot ends places a capstone on one of the novel's most striking series of motifs, which highlight animation, automatism, and the examined body. From the mad veteran's declaration that the invisible man is a "walking zombie . . . [a] mechanical man!,"[14] through a series of electrified and apparently dancing black bodies and mechanical automaton dolls, to this perplexing final walking bridge, the novel's uncanny and discomfiting images suggest that automatism is in fact the novel's central concern. Ellison uses the automaton figure in an innovative way, satirizing and critiquing the political consequences of scientific discourses that were central to a new approach to race in the postwar United States. When he poses his protagonist as the automaton and then dissects the gazes of other characters upon the figure, Ellison leverages an ethical dimension of perception throughout his major novel, what Stanley Cavell identifies as a tension between knowledge and acknowledgment.[15] This ethics of perception has been largely overlooked by the rich exegeses of the text that have focused on the African American cultural heritage and linguistic strategies that Ellison also embraces, just as it has been by admirable studies that have connected Ellison with a variety of important mid-century discourses.[16] I read instead the dynamics of Ellison's anti-institutional *ethos*, which also gives way to how Jacques Rancière might describe its politics: a postwar situation in which scientific discourses purport to "count" African American voices as part of a democratic community and yet in fact place those voices under erasure.[17]

The narrator of *Invisible Man* has his closest brush with a scientific approach to race in what the novel calls the "factory hospital," after he has been in a traumatic accident working in the paint factory of which the hospital is a part. Like Kesey with his "Combine" and Ginsberg with "Moloch," Ellison combines institutions as a strategy for underlining their similarities and pervasive control over U.S. culture. The invisible man lies strapped to a bed, as the doctors standing above him hold a long discussion about his "case."[18] While the lab-coated, shock-administering doctors would seem at first glance to be psychologists, there are several reasons to see sociologists as the object of Ellison's satire in this pivotal scene. In the discussion, one doctor asserts, "It would be more scientific to try to define the case. It has been developing some three hundred years."[19] This statement signals that the scientists are speaking about the narrator not as an individual but rather as a product of the historical forces that have conditioned and determined him, suggesting the sociological mode of analysis that came to dominate discussions of the race question in the mid-century period. Further indicating that sociologists are the intended target of Ellison's satire in this scene, an early draft of the novel has one confused character identify the scientists as "socialists, sociolosts, sociologists? I don't know."[20] This slippage between a setting that belongs to psychology and a discussion that pertains to social structures and politics signals that this scene's satire reaches beyond later antipsychiatry novelists' jibes at the mental institution as a site of normalizing discipline. This duality in the scene indicates that the dynamics of automatism and misinterpretation surrounding the shock treatment contains instead a suggestive response to the epistemological limitations of midcentury sociological approaches to race.[21]

While the sociology of Robert Park and the Chicago School more generally has been discussed at length with regard to the intersections of race, gender, and sexuality in *Invisible Man*, the most significant sociological text for understanding the novel's approach to politics, Gunnar Myrdal's *An American Dilemma*, has yet to be fully explored.[22] *An American Dilemma* set many of the terms of debate and horizons of success for the civil rights struggle, and it was notably the first white-authored study not to assume outright the inferiority of African Americans.[23] This multi-year study was prompted by the 1935 Harlem riots,[24] which provoked the Carnegie Foundation to fund an investigation into the "Negro Problem and Modern Democracy," to be led by Myrdal with research assistance from a great many African American sociologists and intellectuals.[25] Myrdal's large-scale sociological study pointed out the contradictions of life in the Jim Crow South, and the titular dilemma is in short the contradiction between the "creed" of American equality and the racist practices of the segregation era. The resulting book, which was released in 1944, pinned its hopes for change on discerning and exposing this contradiction.[26] Although Ellison found points worthy of praise in his then unpublished 1944 review of *An American Dilemma,* he finds in the main that African Americans continue to play the role of the object in the study; they are, through the very methods used to confront the problem, denied any measure of autonomy or agency in this descriptive text. According to Ellison, Myrdal concludes that "the Negro's . . . opinions on the Negro problem are, in the main, to be considered as secondary reactions to more primary pressures from the side of the dominant white majority," suggesting that, at best, African Americans' ideas are a direct and unreflective effect of social inequality.[27] This charge is borne out through the study: even as Myrdal troubles to cite a wide range of African American intellectuals, including W. E. B. Du Bois, Richard Wright, and Ralph Bunche, the text gives little credence to what it refers to as "Negro popular theories," defined as "consciously thought-out, though not necessarily logical or accurate system[s] of ideas held by a large group of people," a rubric under which he places the spectrum of African American political positions from NAACP activism to Garveyism.[28]

Ellison's review offers a powerful rejoinder to Myrdal's statement about political ideas as "secondary reactions": "can a people . . . live and develop for over three hundred years simply by *reacting?*"[29] Ellison's criticism implies that the methods employed by even a sympathetic sociologist like Myrdal can fail, in the main, to understand such an important aspect of the problem as more than a kind of reaction formation. Through the absurd image of the Negro making his "way of life on the horns of the white man's dilemma," Ellison suggests that the African American's existence *for whites*—that is, his reaction to conditions imposed by a white power structure—seems to him to constitute the extent of a white sociology's knowledge of the Negro.

More recent Marxist criticisms of Myrdal have taken a different tack, suggesting a different mode of corrective to ideologies of postwar racial liberalism. In *Race and the Making of American Liberalism,* Carol Horton draws attention to the "structural, class-rooted developments [that] were simply off the radar screen of the postwar liberal mind."[30] Horton claims that "postwar liberalism contained an internal contradiction that would sabotage its credibility and potential" by posing discrimination alone as the main barrier to national progress.[31] And in line with Horton's thinking, Ellison does sharply criticize Myrdal's "running battle with Marxism" and the absence of "class struggle" in the study's analysis, and he even wonders aloud whether the ultimate goal of such a study is the *"more efficient exploitation of the natural and human resources of the South,"* such that the collection of scientific data is a tool of class domination.[32] Though the economic and "structural" elements of the race question do not exactly constitute a blind spot for Ellison's review, the novel's approach to anticommunism leads him to put his faith in cultural and creative forms of political action that might be able to bring about new forms of recognition for African Americans. Such recognition functions as a prerequisite, for Ellison, to the kinds of structural analysis and redress that Horton rightly deems as urgent and important facets of racial politics.

Despite its impassioned and suggestive objections to Myrdal's methodology, Ellison's review article on Myrdal's *An American Dilemma* would not be published until 1964, when it was collected in *Shadow and Act.* The essay had been rejected from its originally intended venue, the *Antioch Review,* as what Ellison called "a mess of loose ends," and it seems, from the essay's somewhat contradictory conclusions, that Ellison had attempted to treat more problems than could be addressed adequately in an essay of its length.[33] In the light of the essay's rejection—which occurred around the same time Ellison turned toward fiction writing for the rest of the decade—it becomes clear that Ellison continues to deal with the same problems in his fiction. Ellison's critique, in the essay, of the sociological approach to race seems difficult to separate, then, from his inclusion of the factory hospital scene in his novel, and particularly the scientist's mention of a "case [that] has been developing for nearly three hundred years."[34] This "three hundred years" echoes provocatively Ellison's charge to Myrdal in his review—"can we develop over three hundred years simply by *reacting*?"—and it likewise strips the narrator of his personal identity by viewing him only in terms of the forces that were supposed to have shaped him.

This discussion of "reaction" as a limiting epistemological perspective develops a critique of a sociological approach to the "Negro problem" along a path that echoes critiques of scientific modernity performed by Ellison's contemporaries. While his discussions of scientific production as a tool of class domination coincide with the main thesis of Adorno and Horkheimer's *Dialectic of Enlightenment,* Ellison holds perhaps a greater affinity with Martin Heidegger's views on science. Heidegger's critique of modern Western science and philosophy in *Being and Time* holds that both enterprises understand "Being" only as substance and that to ask the question of Being is to recognize different modes of being: in addition to being as substance, there is also the being of equipment and, most important, the being of *Dasein,* or human existence. The latter is the site of culture (what Heidegger calls inhabiting a "world"), of anxiety and care, and of intersubjective connection.[35] In his seminal essay on scientific modernity, "The Age of the World-Picture," Heidegger argues that the assumption that the world can be measured as substance alone has dominated the worldview of our era, an observation that proves particularly relevant to scientific projects like Myrdal's. The project of science, for Heidegger, is to create and manipulate an ever more accurate picture of the world as substance: the ideal of science is to "set up nature to exhibit itself as a coherence of forces calculable in advance."[36] To be "calculable in advance" is to be treated as a simple substance, and Ellison's essay on Myrdal seems to make precisely this point, that Myrdal approaches African Americans as exhibiting a "calculable" range of reactions.[37] Although scientific studies proved to great advantage in court cases like *Brown v. Board,* they could also have negative effects. The observation and prediction of behaviors (as in behaviorist psychology); the statistical analysis of the relationships between family structure, income, and mental illness; and overly deterministic views of the shaping power of institutions all contribute to a reductive vision of the subjects under question, what Mark Seltzer has provocatively called "statistical persons."[38] Likewise, the narrator, as a "case that has been developing three hundred years," is not a democratic subject but rather an object to be measured from a distance, a problem to be solved through the one-sided administration of a cure.

The factory hospital scene in *Invisible Man* will ultimately condense the dynamic of action and reaction discussed in the *An American Dilemma* review into an examination of the protagonist's body under jolts of electricity. When the scientists in the factory hospital ultimately give the narrator the shock treatment, they identify his jitters and hops along with the electric current as dancing. "Get hot," they tell him, saying, "they really do have rhythm, don't they?"[39] This moment plays within a similar scene of interpretation, as these scientists pretend to postulate a cultural observation, in the form of a stereotype, in what ought only to be a physiological reaction to the electric current. The exclamation that "they really do have rhythm" ties a caricature of the narrator's culture to a physiological reaction, and the scientists in the factory hospital inhabit the same short-circuit through which African American culture and political views can be understood as pathologies. This pattern of simple action and reaction bears comparison to the automaton bank the narrator finds later in the novel.[40] As the narrator describes its operation, "if a coin is placed in the hand and a lever pressed upon the back, it will raise its arm and flip the coin into the grinning mouth."[41] By making the narrator dance at the flip of a switch, the ECT-administering doctors are using the narrator's shock treatment for entertainment in the same way a child might use such a bank for a moment of laughter.[42] That this application of electricity causes laughter in the doctors and likely a feeling of mild horror in the reader recalls the range of aesthetic reactions to automatons that Henri Bergson and Sigmund Freud had explored. Freud pointed to realistic automatons as a source of the feeling of the "uncanny," whereas Bergson had identified humans' resemblance to machines as the wellspring of all comedy. This range of possibilities points once more to a deep-seated confusion that automatons provoke, especially when the decision to treat something as a person or as a thing carries profound ethical consequences. That the doctors do not count the narrator as fully human amplifies, in the register of satire, Ellison's critique of Myrdal.

With the factory hospital scene, as Ellison chose to shape his fiction, the treatment seems not, ultimately, to have the effect those scientists desired, namely, that "society [should] suffer no traumata on his account."[43] The narrator becomes, despite the treatment, something of a troublemaker for the rest of the novel, rather than being calmed and neutralized, put happily out of sight and mind from a white America that might wish him out of existence. The lasting consequence of this scene, for the purpose of the novel's plot, is that the narrator forgets his name: a result that exemplifies the doctors' decision to treat him as a mere symptom of American history.

A strong continuity holds between Ellison's figurations of automatons in his critiques of scientific management of the race question and the metaphors that surround his participation in a "scientific" Communist Party, one that suggests a reframing of current debates about the novel's anticommunism. In asserting that the main object of critique is the party's scientism, Ellison's anticommunism itself takes on a less important role in our understanding of the novel.[44] Ellison's habit of continuing to favor Marxist modes of analysis, even after his break with the Communist Party in 1943, shares more in common with postwar Trotskyites such as Daniel Bell, Norman Mailer, and Irving Howe, whom Andrew Ross characterizes as "protective of what they saw as the privilege of artistic 'freedom' over and against political 'discipline,' temperamentally unsuited to the steadily committed life of the organized 'professional revolutionary.'"[45] The drafts of *Invisible Man* suggest, however, that whatever Ellison's "temperament," he wanted to radically alter the Communist Party's approach to political change, an approach that had been tainted by a technocratic ideology that elsewhere pervaded the culture of the 1940s and 1950s. As I described in chapter 1, intellectuals who defined American identity against the concept of "totalitarianism" often figured the scientific state control of society as the primary link between a eugenic fascism and a scientifically Marxist communism.[46] The successful implementation of wartime cinematic propaganda, in the United States and Nazi Germany, and the popularity of novels like Aldous Huxley's *Brave New World*[47] and George Orwell's *1984* in the United States all made Americans newly anxious about the prospect of what Theodor Adorno would call the "administered world."[48] Ellison's automatons, however, would focus this image, not on the question of an inscrutable Communist Other, but on the Other that the African American would present to the white American expert.

Invisible Man depicts its protagonist's initiation into the communist Brotherhood as a process of learning a "scientific terminology" and "speaking as a scientist."[49] Even the Brotherhood's abandonment of the Harlem district is deemed a "scientific necessity," and the novel makes approximately two dozen other mentions of science and scientists in connection with the Brotherhood.[50] At the moment of his greatest infatuation with the Brotherhood, the narrator describes the situation: "it was a world that could be controlled by science, and the Brotherhood had both science and history under control. . . . We recognized no loose ends, everything could be controlled by our science. Life was all pattern and discipline."[51]

This encapsulation of the party line presents the narrator's embrace of the party as tied closely to control, mastery, and objective certainty: "everything could be controlled by our science." Indeed, the Brotherhood, in introducing the narrator to Marxism, tells him to forget his economics and his sociology that he learned at the university—not because, as the narrator might hope, he will not need them anymore but because the party's Marxism provides an even more precise, and more complete, objective picture of the world.[52] Taken together with the infatuation with discipline, the narrator's participation in the party, as "pattern and discipline," takes on an automatic quality: it becomes another mode of objectification, and the party's potential for collective action is undermined by a hierarchical, secretive, and capricious managing class.

Of course, anxiety about party discipline would indeed have tapped into a large set of anxieties in the postwar United States: the dialectic between behaviorist and deterministic theories of the self and representations of totalitarian others as automatons was, at this point, reaching a point of crisis. Andrew Hoberek has astutely noted that Ellison's Brotherhood would also have evoked the cultural discourse of the "organization man," an anxiety of the newly enlarged post–World War II professional–managerial class.[53] This anticonformity discourse arguably became one of the defining features of the literature of the 1950s, and a strong point of continuity with the literatures of the 1960s, which often expressed similar sentiments about conformity in more outrageous ways. Numerous discourses of American identity in the 1950s bear the marks of this new international situation, and we can read the "organization man" discourse as actively excluding the aspects of American culture that might be seen to resemble the mob psychology and the perceived conformity that made totalitarianism possible, as *The Manchurian Candidate* would do a decade later. In this sense, *Invisible Man* partakes in both the "organization man" discourse and the new antifascist and anticommunist sentiment in the United States at the time. All three of these contexts are evoked in the dramatic irony of the narrator's infatuation with discipline and in the Brotherhood's instrumental treatment of the protagonist.

This connection with an "organization man" discourse, though convincing through much of the novel, loses some of its explanatory power when the Brotherhood's discussions turn to the question of race. A professed antiracism is central to Brother Jack's character in the novel, and this antiracism draws the narrator toward the party's "real democracy."[54] When, for instance, the Brothers ask the narrator to sing, because, they say, "all colored people can sing," Jack attempts to silence them and becomes angry at the instance of stereotyping.[55] This moment marks Ellison's satire of him as an explicitly antiracist avatar of a new racial liberalism, even though the narrator will at one point attempt to reduce Jack to the other white racist figures he encounters (when he suggests sardonically that he be called "Marse Jack"[56]). When the Brotherhood "sacrifices" the Harlem district, inciting and then abandoning a riot, the narrator sees a problem that persists despite this professed antiracism: he asks, "What did [the Brotherhood] know of us [African Americans in

Harlem], except that we numbered so many, worked on certain jobs, offered so many votes, and provided so many marchers for some protest parade of theirs?"[57] In this moment, the party's instrumental treatment of the Harlem district—as a means to an end rather than an end in itself—takes the form of the count, and they are numbered only as manipulable bodies. Suggesting Heidegger's notion of a "world-picture," the narrator claims the Brotherhood was "all a swindle, an obscene swindle! They had set themselves up to describe the world," which is to say that they had cultivated an objective distance from history to try to manipulate it in a "scientific" manner.[58]

The political consequences of the party's objectification of African Americans are most strongly expressed in a conversation between Jack and the narrator that would be cut from the final version of the novel:

> "Look," I said, "but aren't my people part of history?"
>
> "Yes and no," [Jack] said. "A people may exist during a historical period and still not be of that period, just as the Indians are still with us but not a part of present day historical movements. . . . They must be able to effectively accept or reject the basic issues of its time, and thus it must learn to act."
>
> "But don't we act?" I said.
>
> "Yes, but not always historically," he said eagerly, "I refer to decisive action. Perhaps it would be more correct to say that the Negro people react rather than act—insofar as they express themselves after those events which profoundly [affect] their destiny have occurred."[59]

In this passage, the difference between being "in" and "of" history is in the final analysis an epistemological difference between ways of seeing and understanding history.[60] Ellison here poses the question of whether agency is an effect of our ways of seeing: seeing someone as "existing during" rather than "being of" a period is a judgment about whether that person has access to forms of power, naturally, but it is also a judgment about where to look for agency. Where the old-fashioned history framed around "great men" has many times undergone criticism for overlooking the role of social movements, one would likely point precisely to a

Marxist history for a lens through which we would "see" a much wider variety of forms of historical agency and collective action. It would seem, then, that it is an overly rigid definition of an organized proletarian class that restricts the Brotherhood's understanding of African American agency. This scientific understanding of historical agency can give us a stronger purchase on the meaning of the novel's repeated mention of "plunging outside history."[61] The "history" outside of which the protagonist plunges is, in the light of this context, the insistently scientific history of a prognosticating dialectical materialism. Moreover, this quotation provides another deep point of continuity between Ellison's critique of the social sciences discussed in conjunction with the novel's factory hospital scene. Ellison's then unpublished review of Gunnar Myrdal's *An American Dilemma* becomes, in this fragment, the basis for the novel's critique of the Brotherhood. Here it is the distinction between "acting" and "reacting" that echoes strongly with the language of that review, which shows that, in Ellison's understanding, it is both the sociologists like Myrdal and the Communist Party, whose limited grasp of African American culture causes them to perceive and represent African Americans as passive bystanders in history, rather than as different kinds of historical agents.

It is this dynamic of interpretation and historical agency that frames the novel's central figuration of automatism, the Sambo doll scene. This scene takes place on the street corner, where the narrator encounters his former colleague in the Brotherhood, Tod Clifton, selling paper and cardboard bouncing dolls called "boogie woogie Sambo." After the narrator recognizes Clifton and tries to approach him, his unlicensed operation happens to be raided by the police, and Clifton is shot after he resists arrest. This suggestive and widely cited scene has been remarked upon for the eerie reciprocal relationship between Clifton and his doll: like the doll, Clifton bounces with his legs, and his face and arms remain stiff, and his staring eyes do not recognize his friend as he makes a repetitive and outlandish sales pitch for the dolls. Previous readings of the scene have noted how the novel uses this uncanny moment to complicate the relationship between African Americans and objects (through a compelling illustration of the Marxian concept of reification) and as a representation of African American "animatedness," but neither of these readings satisfactorily addresses the scene's function for the plot or its wider significance for the novel.[62]

Clifton's importance in the earlier scenes of the novel has to do with his susceptibility to other ideologies. Ras the Exhorter, the novel's parodically reductive representative of black nationalism, seeks out Clifton, and his susceptibility to Ras's seduction takes the form of a locked gaze: after a dose of rhetoric from Ras, "Clifton looked at Ras with a tight, fascinated expression, pulling away from" the narrator.[63] The possibility that Clifton could be swayed so easily from a party Marxist position to a Black Nationalist one, then, evokes a totalitarian conformity, the possibility of becoming a mass man. Later, by selling these dolls in the street, Clifton participates in his own denigration in a profoundly unsettling way, suggesting that he might be susceptible to any kind of mental manipulation. It is after the encounter with Ras that Clifton suggests that he might need to "plunge outside of history," and the Sambo doll episode takes on an explicitly political valence when the narrator reconsiders this earlier statement. He judges that Clifton had "fallen outside of history" but is made uneasy, thinking Clifton "knew that only in the Brotherhood could we make ourselves known, could we avoid being empty Sambo dolls."[64] This metaphor of emptiness suggests that the narrator believes what Jack has said about agency at this point: that only within the white-organized proletarian class can African Americans act "historically."[65]

When the narrator encounters Clifton selling Sambo dolls on the street, he finds a dilemma in which he suspects that Clifton has, indeed, ceased being a rational agent, and the passages in which this encounter is described produce a singular and strange effect. Between the narrator, the doll, and Clifton, it seems as though the doll's automatic movements infect the other two characters. First, the narrator describes the Sambo,

> a grinning doll of orange-and-black tissue paper . . . [that moved] up and down in a loose-jointed, shoulder-shaking, infuriatingly sensuous motion, a dance that was completely detached from the black, mask-like face. . . . It was Clifton, riding easily back and forth in his knees, flexing his legs without shifting his feet, his right shoulder raised at an

angle and his arm pointing stiffly at the bouncing doll as he spieled from the corner of his mouth.[66]

Sianne Ngai has discussed how, in the language of this passage, Clifton and the doll are co-implicated in the doll's movement, such that "the human agent anthropomorphizes the puppet . . . but the puppet also mechanizes the human."[67] This contrast, the detachment, between the easy and "sensuous" rocking of both figures simultaneously and the stiffness of their expressions signals that either Clifton is imitating the doll or that, in playing the part of the spieling salesman, he enacts a form of entertainment much like that of the doll itself. The narrator, in the moment of recognizing Clifton, describes himself as "paralyzed" before he moves to spit on the bouncing doll.[68] This is clearly an uncanny moment, the similarity between Clifton and his Sambo doll's movements producing an uncomfortable confusion for the narrator, which leads in turn to his own object-like paralysis. To call once again on Bergson's interpretation of the automaton, it seems particularly strange that the Sambo doll should produce such a great deal of *laughter* in this scene: *"He'll keep you entertained. He'll make you weep sweet— / Tears from laughing,"* claims Clifton, and the crowd around Clifton continues laughing throughout the entire scene.[69] The laughter is perhaps the scene's most nonsensical element—it seems difficult to imagine that a cardboard doll, even if it does move "as though it receive[s] a perverse pleasure from its motions," could captivate a crowd so completely.[70] Rather, it is the proximity between Clifton's movements and those of the Sambo doll that gives the viewers the impression that the seller and the product are two halves of an entertaining vaudeville act. When the narrator spits on the doll, his attempt to refuse the spectacle actually implicates him in it, as he sees "a short pot-bellied man look down [at the doll], then up at me with amazement and explode with laughter, pointing from me to the doll."[71] The man apparently laughs because he supposes the narrator would be dim-witted enough to mistake the Sambo doll for another black man, as in the stock situation of comedy routines that can be traced back to Joel Chandler Harris's "Wonderful Tar-Baby" story. The difference, however, between the Tar-Baby and the Sambo doll scene is that the latter is both a comic and tragic encounter with a sham interlocutor: for the white man who explodes with laughter, the scene is a comic one, but the scene's uncanny force "paralyze[s]" the narrator, and the scene ultimately ends in Clifton's tragic death at the hands of the police.

The episodes with Clifton echo and reframe Ellison's critique of Myrdal and the factory hospital scientists, particularly when Clifton's death sets in motion the novel's final events, including the narrator's break from the scientific Brotherhood. In what is perhaps the central ethical gesture of the novel, the narrator publicly mourns Clifton and defends his actions against the Brotherhood, who calls Clifton a "traitor."[72] The Brotherhood has used Clifton as a means to an end, discarding him when he is no longer valuable; by contrast, the narrator defends Clifton as "a man and a Negro; a man and brother [even though he was] jam-full of contradictions."[73] the moment when the narrator finds Clifton on the corner, he spits on one of the dolls to signify his disapproval of Clifton's actions, but he nevertheless insists on trying to reach Clifton, even though he had apparently lost his conscious autonomy and refused to return the narrator's gaze. Despite his uncertainty regarding the humanity of his former friend, the narrator makes a leap of faith in pursuing him and in defending his proper burial. As with the novel's other interpretive dilemmas involving automatism, an act of acknowledgment across a gulf of uncertainty becomes the pivotal ethical gesture.

Unlike the factory hospital scientists, who were content to treat the narrator as a machine unworthy of ethical obligation, the narrator here offers acknowledgment despite his uncertainty about Clifton. That this acknowledgment brings about the final break from the scientific Brotherhood can remind us again of the general tenor of the narrator's party involvement. The main source of tension in this involvement is that, in a series of tests and orations, the narrator is not sufficiently "scientific" or "theoretical" in his approach—in his final argument with the Brothers, they call him, sarcastically, a "great tactician" and "quite a theoretician."[74] His job, the Brotherhood tells him in this final argument, is to "keep repeating the last thing we told you to say," suggesting that, as an organization man, he is acknowledged as an object, and as a tool, but not as a rational individual capable of thought and action.[75]

Where Brother Jack states that "the Negro people react rather than act," we find that this excised conversation provides the proper context for a frequently cited moment in the novel's riot scene, in which the narrators sees Harlem's poor burning their tenement buildings. The narrator declares that, despite what the Brotherhood has claimed, these Harlem residents are "capable of their own action."[76] It is ultimately in the capability for action and in creative political gestures that Ellison's novel finds hope, in a small set of affirmative answers to the novel's well-known question, "can politics ever be an expression of love?"[77] The liberation from the tenement buildings in the riot as well as the narrator's musings on the zoot-suiters—"who knew but that they were the saviors, the true leaders, the bearers of something precious?"—are two well-rehearsed examples of this creative expression.[78] Less often noted, however, is the antieviction parade that the narrator puts on early in the novel, in which he employs a kind of dancing or step team, the "Hot Foot Squad," which delights the crowd and "dumbfounds" the police, a mood of protest that anticipates the creative political culture of the 1960s.[79] On the level of the novel's language, there is for instance the exchange between the black vet and the white donor Norton in the Golden Day bar, in which the vet had been "trying to change some blood into money," and the vet "discovered it and John D. Rockefeller stole the formula from [him]," a brilliant mix of Marxian thought about the nature of labor, celebratory nod to Charles Drew (the African American inventor of blood plasma), and veiled criticism of white philanthropy's agendas, as Rockefeller had been one of Booker T. Washington's benefactors.[80] In addition to the novel's language and reported actions, such as the parade, the novel ultimately showcases the narrator's orations as a creative political action. This oratorical skill, which leads to his underground hibernation and the scene from which he writes, is not far removed from the form of the novel itself as a form of creative politics: the dialectical structure of the fiction is itself a testament to the paradox, contradiction, and depth of experience that remain invisible to the technocratic gaze.

Ellison's novel, then, is perhaps most concerned with articulating an ethics of acknowledgment implicit in the perceptual problem of the automaton. In his 1953 National Book Award acceptance speech, Ellison claims the novel attempts to return to "the mood of personal moral responsibility for democracy."[81] This seems like a somewhat conservative statement on the face of it; it might be read, in the light of his disavowals of Wright's "narrow naturalism" later in the same speech, with an emphasis on personal responsibility, a rejection of rhetorics of damage or blame. Yet the question of a responsibility for democracy might persuade us to read *Invisible Man* as an exploration of acknowledgment, the acknowledgment of others as rational and capable political agents who must be the sine qua non of democracy. This is the acknowledgment withheld in the scientific management and legal disenfranchisement of an African American population, and in a scientific communist party's instrumental treatment of a community, even as perpetrated by staunchly antiracist individuals. This acknowledgment, moreover, is not a mystical or elusive element: it consists in acting as if a particular subject is capable of playing an active role relative to her surroundings. Ellison's novel brings out the dynamics of this acknowledgment through the parodically reductive image of the African American automaton, urging the reader to recognize anew the complexity and irreducibility of human experience.

Anonymous Biological Robots in Feminism, Progressivism, and Consumer Culture

The direct echoes of Ellison's scenes can be seen in both the novel and film of *One Flew over the Cuckoo's Nest,* as well as in African American texts such as Ishmael Reed's *Mumbo Jumbo,* in which a black "Talking Android" resembles the naive Invisible Man in parroting white ideologies, and even in Janelle Monáe's series of albums that began in 2008, in which a black android escapes within a futuristic city named Metropolis.[82] But its indirect echoes, and echoes in turn of World War II–era antitotalitarian automaton rhetoric more generally, could be found in a wide variety of political pronouncements in the 1960s and 1970s. Take, for instance, Charles A. Reich's epochal *The Greening of America,* which gives a particularly forceful and lengthy adaptation of automaton imagery to postwar U.S. political life:

What kind of life does man live under the domination of the Corporate State? It is the life that was foreseen in *The Cabinet of Dr. Caligari, Metropolis,* and *M,* a robot life, in which man is deprived of his own being, and he becomes instead a mere role, occupation or function. The self within him is killed, and he walks through the remainder of his days mindless and lifeless, the inmate and instrument of a machine world.

The process by which man is deprived of his self begins with his institutionalized training in public school for a place in the machinery of the State. The object of the training is not merely to teach him how to perform some specific function, it is to make him become that function; to see and judge himself in terms of functions, and to abandon any aspect of self, thinking, questioning, feeling, loving, that has no utility for either production or consumption in the Corporate State. The training for the role of consumer is just as important as the training for a job, and at least equally significant for the loss of self.[83]

These opening paragraphs to Reich's chapter on "The Lost Self" recapitulate in miniature a history of the human automaton, as he visibly adapts the machine-age image in the first paragraph to the institutions—combined into the "Corporate State" just like the Combine and factory hospital—of the postwar United States by way of a language of "lost self" borrowed from Erich Fromm's antitotalitarianism. In this adaptability, the automaton could become the Plato's Cave of progressive rhetoric, in which false consciousness of consumerism or patriarchy or racist ideology or obedience to authority itself could be emblematized in the unthinking human automaton. We can see the range of this flexibility at its extremes in Betty Friedan's feminist rhetoric in the 1960s and, in turn, its commercial adaptability in a particularly illustrative advertisement from the 1980s.

As Reich's emphasis on consumption indicates, the automaton image would prove useful for criticizing consumer culture and advertising. Vance Packard's *The Hidden Persuaders* had introduced the subliminal message as a notion that made the range of corporate advertising practices seem all the more manipulative, and consumers all the more passive and unthinking.[84] Feminist thinkers of the 1960s and 1970s framed consumption as an explicitly feminist problem, and Friedan framed the consumerist patriarchy as a combined institution as a rhetorical strategy for fighting against it. In *The Feminine Mystique,* Friedan interviews an ad man who tells her that "American housewives can be given the sense of identity, purpose, creativity, the self-realization, even the sexual joy they lack—by the buying of things."[85] The ad man's description exactly inverts Friedan's argument in *The Feminine Mystique,* which encourages women to see through false versions of "identity [and] purpose," and the rest to find their own forms of agency. In this well-known line of argument on the "Sexual Sell," Friedan details many of the advertising strategies that leverage housewives' insecurities and even their feelings of "guilt" for having free time.[86] Advertisers for cake mixes, in another example, "manipulate [the housewife's] need for a 'feeling of creativeness' into the buying of his product."[87] The programs of consumer culture, in Friedan's account, presuppose a female subject who is readily manipulable, gullible, and insecure, a subject whom Friedan illustrates with images of automatons and totalitarianism.

A rhetorically masterful book, *The Feminine Mystique* links the false consciousness of the consumer–housewife to a wide variety of images, ranging from Korean War POWs to broken-down concentration camp inmates to unthinking machines. In one of the text's most striking images, the housewife "turns away from individual identity to become an anonymous biological robot in a docile mass. She becomes less than human."[88] By way of this comparison to a robot, Friedan leverages cinematic and literary imagery from science fiction as a way to imagine the housewife's lack of self-actualization. Moreover, with the phrase "docile mass," she sets up a set of extended comparisons between the institutions that massify individuals (as Hannah Arendt had described it in the totalitarian transition from "classes" to "masses"[89]) and the household itself, which appears to be the furthest thing from an institution.[90] As I have described this institutional automaton rhetoric with Kesey and Ellison, Friedan here asks her readers to imagine society's relatively open systems as total institutions, so that what seem like ordinary social interactions can be comprehended as insidious institutional relationships.

When the household becomes a "comfortable concentration camp," as Friedan calls it, it is first and foremost the site of lost autonomy.[91]

Friedan goes on to strengthen this linkage between her science fiction–esque "biological robot" imagery and the automatons of World War II and Korean War discourse to shocking effect. After recounting studies of the stunted psychology that results from housewives' "progressive dehumanization," she points to what she refers to as an "uncanny, uncomfortable insight" into that psychology, namely, its similarity to that of prisoners in Nazi concentration camps.[92] Recognizing explicitly the aesthetic impact of this comparison, Friedan cites postwar thinkers including David Riesman to juxtapose psychological studies of housewives, POWs, and concentration camp inmates.[93] The managerial strategy of progressive dehumanization that gives the chapter its name links these victims as captives and, most important, as a depressive, broken-down variant on the human automaton: the human reduced to bare life, to an organism without volition or autonomy, a stark and affecting contrast from advertisements' and television's conventional images of the cheerful housewife.[94] For Friedan, the housewife's false consciousness is homologous to that of those who "adjusted to the conditions of the camps [and] surrendered their human identity and went almost indifferently to their deaths."[95] Friedan here emphasizes the "dehumanization" of the housewife's situation as depriving women of the minimum sufficient conditions for human self-actualization and "identity." There are few concrete illustrations of such "identity" in *The Feminine Mystique*; it is defined instead primarily in the negative terms of the concentration camp's sublime unfreedom. The chapter "Progressive Dehumanization: The Comfortable Concentration Camp" ends with a parable for fighting back, moreover, borrowed from Bruno Bettelheim. It tells of a group of naked concentration camp prisoners who were "no longer human, merely docile robots," and who "lined up to enter the gas chamber."[96] A guard hears that a prisoner had formerly been a dancer and asks her to dance. "She did, and as she danced, she approached him, seized his gun and shot him down." Bettelheim's observation, which Friedan cites, is that in being recognized as a dancer, "no longer was she a number, a nameless depersonalized prisoner, but the dancer she used to be."[97] Friedan concludes her parable by exhorting women to "exercise their human freedom, and recapture their sense of self . . . [to] refuse to be nameless, depersonalized, manipulated." They can imagine their way out of the total institution. It is when the individual believes in the captor's or patriarch's reduction of her to a mere object that she *becomes* less than human. For Friedan, the solution is to educate women, to persuade them to reject the ideology of the feminine mystique and of the sexual sell. And as with the false consciousness of fascism that Erich Fromm identified during World War II, the main problem in the home is a psychological one, and "the most important battle" for Friedan "*can* be fought in the mind and spirit of woman herself."[98] The plan for casting off this false consciousness revolves around seeing the labor that women perform in a different way. Accordingly, in the conclusion, Friedan states, "The first step in [the] plan [for a new, whole life as a woman] is to see housework for what it is—not a career, but something that must be done as quickly and efficiently as possible."[99] In this way, Friedan imagines that the woman's demand for acknowledgment and even just her embrace of individual freedom will catalyze the solution to the problem that has no name.

Friedan's dynamic of household control is recast, both eerily and parodically, in Ira Levin's *The Stepford Wives* and Bryan Forbes's 1975 film adaptation of the novel starring Katharine Ross.[100] This remediation of Friedan's captive housewife image sets aside the concentration camp and POW imagery of Friedan's work but amplifies and literalizes Friedan's several comparisons of the housewife to a "robot" who exists to fulfill others' desires. The protagonist, Joanna Eberhart, moves to the suburbs and almost immediately begins, in a fashion that Friedan would approve of, to start "consciousness-raising" groups among the town's women, who seem all too eager to please their husbands. In those meetings, the other women trade housekeeping tips and tricks, and they parrot cleaning product ads. Soon thereafter, Joanna stumbles onto the town's conspiracy, that these women are animatronic robots, built by their engineer husbands to replace actual women. Levin's cruel twist in the novel is to give Friedan herself part of the blame for the trouble: Joanna finds an old newspaper article about Friedan's inspirational visit to Stepford, in which "over fifty women applauded Mrs. Friedan . . . as she cited the inequities and frustrations besetting the modern-day house-wife."[101] In a directly reactionary gesture, the husbands have instated Stepford as a town of robots, a nostalgic stronghold itself inspired by a resentment of feminist activism.

In terms of the confluence between automatism and totalizing social space, Levin and Forbes borrow Friedan's rhetoric of the total institution by making the exclusive suburb of Stepford into a closed-off space of total control. And it's this total control that constitutes the husbands' fantasy of the suburbs, a widely shared ideal of controlled environments, homeownership, safety, and ordered households.[102] Here the techno-crats' desire for malleable subjects takes its most extreme form in their literal replacement by robots. Once these multiple resonances are estab-lished in Forbes's film version, the film can rely entirely on campy scenes of automatism for its visual suspense. One robot wife malfunctions early on in a harmless, repetitive fashion that clues in the viewer to the mechanism, and in the film's climax, the characters who had been sym-pathetic protagonists are shown replaced by their uncanny robotic dou-bles and played for a catalog of effects. Against the earnestness of Ralph Ellison's ethical appeal to acknowledgment in his ECT and Sambo doll scenes in *Invisible Man*, *The Stepford Wives* plays the humorous and un-canny aesthetic effects of the automaton as pure genre, as camp horror in the lineage of *Invasion of the Body Snatchers* (Don Siegel, 1956). The feminist arguments that Joanna articulates in the text are undercut, if only a little, by a not-altogether-sympathetic portrayal of her character and through its depiction of the women's liberation movement as an antagonistic enterprise. Friedan called the book and film a "rip-off," and feminist scholars and mainstream media outlets have had lively debates about the text's gender politics, which seem to evade certainty under lay-ers of irony.[103] Instead, the text's most stable bedrock is its narrative of the individual railing against the conspiracy and the total institution before succumbing as its total automaton, yet another variation on *1984*.

To conclude this chapter's sketch of the automaton's adaptability for agendas including antiracism, antipsychiatry, and critiques of the patri-archy and consumerism, it seems necessary, too, to cast a glance at its cooptation by commercial culture itself. If 1950s and 1960s progressives reappropriated the antitotalitarian automaton image to rail against echoes of totalitarianism they saw in the combined-and-therefore-total institutions of "the system," then by the 1970s and 1980s, that anti-institutional *ethos* was in turn ripe for cooptation by commercial culture and even neoconservative politics. An emblematic moment of this commercial reappropriation is Apple's famous "1984" Super Bowl advertisement for the Macintosh computer. For Apple's advertisement to be legible as an expression of Silicon Valley cool, it is necessary to read its use of Orwell in the light of the transformations that the human automaton figure had undergone in the 1960s and 1970s.

The Apple Corporation advertised its new graphical user interface (GUI) computer not by depicting the computer itself or describing its features but in an instance of pure branding, the association of a feeling with the product.[104] And that feeling is something like freedom, with inflections toward coolness, progress, and rebellion, what Thomas Frank would describe in terms of computers being marketed as "devices of liberation."[105] The advertisement, directed by Ridley Scott, begins with images of bald proletarians, reminiscent of the opening scenes from *Metropolis,* marching into a theater, intercut with a young woman in jogging gear, running with a sledgehammer. These shots are then intercut with an enormous face on the screen in the theater, the bald proletarians wearing strange glasses to watch him. Throughout, the face on screen describes in triumphant terms a "garden of pure ideology where each worker may bloom"—a play on Chairman Mao's thousand flowers—and a "unification of thoughts" that eliminates disagreement, before the jogging woman throws the sledgehammer through the screen. The ad refers directly to Orwell's *1984,* and its staging of that novel's "two minutes' hate" also features a healthy dose of *A Clockwork Orange*'s theater-brainwashing scene, from Anthony Burgess's 1962 U.K.-based, anti-welfare-state version of the anti-institutional automaton narrative, as well as Stanley Kubrick's 1971 film adaptation of it.

As branding, Apple's image need not make any sort of literal sense: a woman throws a hammer at a screen as . . . an expression of her individualism, or the rebellion that symbolizes bringing down the totalitarian

The Apple Corporation's "1984" advertisement for the Macintosh personal computer, first aired for the 1984 Super Bowl. The jogging woman, an embodiment of freedom associated with the product's brand, uses the hammer to destroy the screen, a gesture of rebellion against a system that has apparently enslaved countless human automatons.

system? The elimination of mainframe computing that was already under way with other brands of personal computer?[106] A revolutionary gesture that is somehow analogous to using a desktop computer with a mouse-driven GUI? The branding for this version of individuality, though, became so iconic that the Apple Corporation remade the ad in 2004, this time with the rebel woman wearing an iPod (to complement the woman's jogging gear). The representational paradigm for representing what is otherwise noumenal freedom works again through simple visual contrast: the jogging woman alone would be an ordinary runner, but, set alongside the masses of doddering human automatons, she is pure individuality. The libertarians and cyberlibertarians of Silicon Valley during this period could hardly wish for a better mascot. That individuality is also crucially "cool," an attitude that had not been part of Orwell's *1984* but a meaning that had accreted on anti-institutional individuals by association with works like *One Flew over the Cuckoo's Nest*, *The Stepford Wives*, and *A Clockwork Orange*.[107] In addition to this anti-institutional *ethos*, the advertisement also associates the sheer inhumanity of the computer itself with totalitarianism, and the following chapter addresses how the computer and cybernetics inspired science fiction to retool the mechanisms and forms of the human automaton.

CHAPTER THREE

Human Programming

Computation, Emotion, and the Posthuman Other

So far, we have followed the figure of the human automaton through the U.S. mainstream political culture of the 1950s and 1960s. It has revealed a story not about the straitlaced 1950s giving way to the cool 1960s but one in which the same narratives of automatic others and human selves were deployed in very different ways, as the human automaton shaped both American descriptions of totalitarianism and progressive descriptions of the domestic establishment. The uncanny, and sometimes the humorous, force of cinematic and fictional automaton imagery became a widespread tool in descriptions of totalitarian governance, the discipline of total institutions like insane asylums, and even, in Friedan's case, the abstract institution of the household. This rhetoric often relied on, in the 1950s, or critiqued, in the 1960s, social scientific categories of knowledge: behaviorist psychology, the psychology and political theory of totalitarianism, and mid-century sociology. Roughly simultaneous with those developments, beginning in the late 1940s, the new paradigm of cybernetics would prompt scientists in many disciplines to rethink their objects of study in terms of systems of communication and control. This new orientation of thought spread quickly to roboticists, biologists, ecologists, computer scientists, and sociologists, and it would also expand the representational possibilities for human automatons. In this chapter, I trace how the computer-inflected idea of human programmability began to circulate in parallel with the uses of automatism that we've seen already.

While the scientific and conceptual developments of cybernetics took place concurrently with those in political culture traced in chapters 1 and 2, their literary and cinematic expressions tended to be cutting-edge rather than mainstream, or emergent rather than dominant, during that period and even in the decades following. Accordingly, experiments with cybernetics were and are largely confined to avant-garde and science fictional texts.[1] Mark McGurl's analytic category of "technomodernism" offers a way to group together novels and films whose interests in science are often tantamount to expertise and that blur the categories of high postmodernism and science fiction into which authors such as Kurt Vonnegut, Philip K. Dick, Thomas Pynchon, James Tiptree Jr., William S. Burroughs, Greg Egan, Neal Stephenson, Kathy Acker, Paolo Bacigalupi, and others interested in cybernetics might fall.[2] These texts often asked questions about the ways that the body and human being might be transformed by cybernetic technology, that is, ways in which the human might give way to something that might warrant being called "posthuman." Moreover, just as Thomas Foster imagines that academic posthumanist theory and science fiction are often two sides of the same coin, this chapter also addresses (and makes use of) posthumanist theory from the 1980s to the present.[3] As I have described it up to this point, a prominent strain of the liberal humanism to which posthumanist theorists often contrast their work had in fact emerged in response to the posthuman threat of totalitarianism's supposed emphasis on behaviorism. Cybernetics affords distinct "posthuman" qualities from behaviorism, but as I describe in this chapter, it proves equally amenable to depictions of freedom and unfreedom, democracy and its enemies.

To pursue our narrative of human automatons as they have been used to describe selves and others, then, we will need to attend to a few of the ways that computation and cybernetics shifted how we see the boundaries between the human and the machine. This shift occurred on an economic register, as a postindustrial economy replaced humans with machines or outsourced labor, as well as in cultural production, where posthumanist fiction and theory began to ask new questions about how technology changes the boundaries between humans, animals, and machines. Across a wide variety of topics—cybernetics, genetics, robotics

science fiction, and cyberpunk—I sketch these descriptions of selves
and others in three ways: first, in discourses about labor, and particularly
devalued forms of labor; second, in the "techno-Orientalism" that updates older ideas about Asians' and Asian Americans' proclivity toward
technology, lack of affect, and inscrutability; and finally, as discourse
about human programming prefigured, in the case of Neal Stephenson's
1992 cyberpunk novel *Snow Crash,* contemporary ways of describing
selves and others in terms of an inclusive democratic multiculturalism
and a fundamentalism whose psychology appears machine-like in its
closure.

Postindustrial Programmability

In 1947, the cybernetics pioneer Norbert Wiener contacted the leadership
of the Congress of Industrial Organizations (CIO) to warn this major
union group of the future consequences cybernetics would bring for
labor politics in the United States.[4] As machines became increasingly
able to perform menial and dangerous tasks in factories, they would
eventually become the most efficient means of accomplishing most
industrial labor. Wiener realized these developments would dramatically alter the face of the workforce and the roles of unions in a changed
postwar economy. Instead of spinning out a science fictitious scenario
or a variation on the John Henry myth, Wiener presented the CIO with a
bottom line: as machines came to replace industrial workers, they would
drive the value of factory labor down to a level where human workers
would no longer be able to compete. Wiener's insights about the future
of industrial labor, and the changing nature of industry in the coming
half of the twentieth century, couldn't make a difference for the labor leaders' strategies at the time. It can serve in retrospect, though, as an early
marker of the United States's transition toward what Daniel Bell called
the postindustrial society, in which knowledge work, affective labor, and
immaterial labor would constitute a much larger portion of the workforce.[5] This encounter brings to the fore a pair of developments in U.S.
culture, cybernetics and the postindustrial economy, that would change
how Americans understood what separates humans from machines.
Human automaton imagery in U.S. culture, particularly in the realm

of science fiction, would adapt to address these linked developments, wherein the nature of labor would shift dramatically with the growth of the professional–managerial class and the service sector, and where the increasing capacities of machines would force Americans to rethink what qualities would be valued as "human," in both philosophical and economic terms.

Six years after Wiener's conversation with the CIO, Kurt Vonnegut made his fiction debut with a novel that takes Wiener's warning as its whole premise, *Player Piano*.[6] In the novel, the best former factory worker, Rudy Hertz, sits in alcoholic idleness with all the former factory workers. Rudy—whose surname connotes the heart in German [*Herz*], hurt, and the measurement of frequency—had been chosen as the worker whose movements would be recorded, in the style of Frank Bunker Gilbreth Sr.'s motion-capture photography. His captured movements were then programmed into machines so that "the essence of Rudy Hertz produce[d] one, ten, a hundred, or a thousand of the shafts."[7] While the 1920s and 1930s visions of automatic human movement in factories— such as those in *Metropolis*, *White Zombie*, and *Modern Times*—depicted the motions of work as formally abstract ideals to which the human body was made to conform, Vonnegut's 1950s vision holds that the movement, by way of recording, will be abstracted away from human bodies altogether, leaving the bodies themselves as a sort of biological relic of an earlier time. This scenario is the nightmare-come-true of those machine-age automaton films of factories, but it would also raise questions about what that postindustrial workforce and even the professional–managerial class would look like, and what its problems would be. As Alan Liu has traced in his history of knowledge work, *The Laws of Cool*, in this transition workers went from being "robots alienated not so much 'man from [fore]man' as man from system," such that a large class of middle managers, too, felt a similar mode of alienation to other workers.[8] This issue structures *Player Piano*'s plot, in which the protagonist Paul Proteus becomes disillusioned with the artificiality of an automated environment just ahead of the advent of the computers, represented by a checkers-playing machine, that will make managers like him, too, obsolete. (This development also derives directly from Wiener's text.) Especially insofar as it details Proteus's rebellion against the suburb and the corporation, Vonnegut's novel works as a version of the anti-institutional narratives described in the previous chapter, wherein institutions make automatons of individuals unless those individuals assert their freedom by rebelling. But Vonnegut also takes up Wiener's warning that machines would change how the advancements of machine technology might change human values.

Key beginnings of cybernetics included Claude Shannon's development of information theory and Norbert Wiener's 1948 coinage of "cybernetics" as the study of "command and control in the machine and the animal." One of Wiener's experiments from 1946 gives a sense of this orientation of thought regarding machines and animals. In this experiment conducted in Mexico City, Wiener and Arturo Rosenblueth examined the relationships between electricity and muscle fibers, using the quadriceps muscles of cats to test how those muscles received messages from the brain.[9] Wiener and Rosenblueth were examining the signal transmission and reception within the individual animal's muscle. Wiener and Rosenblueth found themselves surprised at the logarithmic curve between current applied to the muscle and its contraction. The muscle itself operated on what appeared like an encoded transmission. As Wiener later put it, "as far as the nervous system works, the individual fibers come very near to showing an 'all or none' action . . . an operation of connected switching extremely like the connected switching of the automatic computing machine."[10] This discovery in biology was roughly simultaneous with Claude Shannon's new theorization of transmission and reception on the Bell telephone network. The telephone network, which Shannon imagined as a single, nationwide computing machine, could be optimized by improving the protocols of information flow. Because cybernetics imagined systems of communication and control in such a flexible framework—at the scales of enormous machines, groups of individuals, individuals, and within individuals—the metaphors of human programming were easily adaptable to that of cybernetic programming.

And indeed, the notion of programming humans developed before and alongside, not after, the notion of programming machines. From

Ivan Pavlov's salivating dogs to John B. Watson's "Little Albert" and
B. F. Skinner's superstitious pigeons, behaviorist psychology had treated
human behaviors and animal behaviors as essentially the same; cybernetics completed a circuit whereby animals, humans, machines, and
networks of all three functioned in essentially similar ways.[11] (William S.
Burroughs would memorably call the human, in the title of a 1961 novel,
The Soft Machine.) In both of his main books on cybernetics, Wiener
would redescribe many preexisting facets of contemporary science in
terms of communication and control, and many academic fields would
be reshaped by this orientation in the coming decades, from Maturana
and Varela's innovations in biology to Niklas Luhmann's work in sociology.[12] In a vivid passage on human psychology, Wiener describes lobotomy and electroconvulsive therapies as analogous to rebooting or wiping
the memory of a computer: "we try . . . to clear the machine of all information, in the hope that when it starts again with different data the difficulty
may not recur."[13] Wiener advocates for great care when treating the body
as one of Burroughs's "soft machines," even as this technocratic gaze
upon the human body would make the instrumentalist reduction of
people to objects a simpler matter. This emphasis on care dovetails in
Wiener's work with an antitotalitarian attitude, which he expresses by
underscoring the differences between human adaptability and the Nazis'
underestimation of that adaptability: the Nazis' "aspiration . . . for a
human state based on the model of the ant [and its predetermined societal roles] results from a profound misapprehension both of the nature
of the ant and of the nature of man. I wish to point out that the very
physical development of the insect conditions it to be an essentially stu-
pid and unlearning individual, cast in a mold which cannot be modified
to any great extent."[14] Indeed, learning and adaptability could stand at
the time as qualities that set humans apart from machines, at the same
time that championing those human qualities could be a way of setting
the democratic United States against fascist authoritarianism.

But the differences between humans and machines within cybernetics could also appear differently, depending on the scale of observation.
In his posthumously published lecture series from 1961, *The Computer
and the Brain,* John von Neumann attempted with the technology from

the time to make the most scientific possible comparison between these apparently dissimilar entities. If both were functionally information processors, and they worked on irreducible units—the neuron and the bit combined with logical operators—then they might fruitfully be compared to precisely qualify their difference in kind. Amid a striking number of functional similarities between brains and computers, Shannon found the most important difference to be that the brain processes primarily in parallel, while the computer processes sequentially.[15] As David Golumbia has described in *The Cultural Logic of Computation*, that image of the mind as computer has also spread through deep assumptions in contemporary linguistics and analytic philosophy.[16] Wendy Hui Kyong Chun has likewise identified a "logic of programmability" that unites genetics, cognitive science, and computer science.[17] Genetics and genomics, however, as they apply to logics of automatism and of human programming, make clear a useful distinction between the programmability of the mind and of the body. Genetics and the discovery of the DNA code developed out of cybernetics' information-oriented approach as it was applied in the 1940s and early 1950s to molecular biology. Lily Kay and others have traced the ways in which the DNA code has been treated as a "book of life," in the sense that the code provides, in writing, the set of biological developments and expressions to which a particular organism is destined.[18] In that sense, the genome programs the body in a way that is analogous to, and usually distinct from, the ways that behavioral conditioning programs the mind. Points of overlap have included moments like the search for the "gay gene" and fictions of genetic predetermination or modification such as Andrew Niccol's *Gattaca* or Octavia Butler's Lilith's Brood series (1987–89).[19] Fictions about genetics often look for a biological substrate to traits such as sexuality, class, or race, interrogating the nature–nurture divide in various ways, though as Jay Clayton observes, they also often focus on an uneasy majority's reactions to new genetic minorities.[20] The texts that track more closely with the issues in *Human Programming* are those in which human programmability makes metaphorical comparisons between the mind and software. These continue the psychological and sociological metaphors of human programmability in terms of new technologies.

Technomodernist texts were relatively quick to adopt this sort of metaphor as cybernetics and computation became mainstays of high-tech corporations and government. In the novel *Nova Express*,[21] William S. Burroughs describes villains, "Nova criminals," who are something like viruses or remote controllers, infecting different human hosts at different times. What "carries over from one human host to another and establishes the identity of the controller . . . is *habit*: idiosyncrasies, vices, food preferences."[22] Such villains make their way through different human subjects, and "some move on junk lines through addicts of the earth, others move on lines of certain sexual practices and so forth."[23] For Burroughs, the distribution, effects, and policing of heroin traffic serve as the basis for an imagined world in which habits, sexual practices, and addictions could be transmitted from person to person. Burroughs describes these "Nova criminals" as a kind of "virus" that is "not three-dimensional," and they feed on "human hosts," which act as "transparent sheets with virus perforations like punch cards passed through the host on the soft machine feeling for a point of intersection."[24] Burroughs imagines the human as programmable—and heroin addiction as a sort of nonhuman state of being—by using a "punch card" metaphor that would then have been on the cutting edge. Rather than using the kind of punch card with which Charlie Chaplin's character clocked in and out of the factory, Burroughs refers here to the punch card used in FORTRAN and other early computer programming languages beginning in the 1950s, each card containing a program or other inputted data. As David Seed and Priscilla Wald have pointed out, Burroughs's metaphor suggests consciousness as a biological virus and as a computer program.[25] Moreover, the programming of information into the mind—as though into a computer— represents a nearly realistic technological substrate for the kinds of propaganda and totalitarian language that Orwell and other earlier figures had described. Here we had a vision of a programmable mind alongside the programmable machines that seemed to function in very similar ways. This horizon of possibility prompted a variety of writers to ask what human functions might still be valued in an age of ever more competent machines.

Emotion and the Mechanical Other

As Wiener's promised revolutions in industry became a reality, a set of feminist thinkers began asking incisive questions about the relationships between labor and value that would reveal, in turn, the kinds of value we associate with emotion and emotional labor. Examining these questions of value can illustrate both the changing boundaries between humans and machines in the period and a new set of strategies for representing machinelike others. In her *Dialectic of Sex: The Case for Feminist Revolution*,[26] Shulamith Firestone considers the political ramifications of a future that Wiener had considered earlier: "cybernation," she writes, the "takeover of work functions by increasingly complex machines," represents a profound change that we will have to "absorb into our traditional value system."[27] The shifting boundary between humans and machines allowed Firestone to reimagine the human by way of negative definition with machines, and she would do so in ways that would anticipate aspects of Donna Haraway's cyborg. The conceptual influences of technology and cybernetics are particularly clear in the feminist writings of Firestone and Valerie Solanas, in which the advent of machines changed their vision of human fulfillment. Solanas and Firestone both developed responses to Betty Friedan's "problem that has no name" in a more radical vein, wherein women might cast off the robotic tasks assigned to them, leaving repetitive and reproductive labor to machines. This notion of reproductive labor, perhaps best formalized as such by Nancy Hartsock in 1983 in "The Feminist Standpoint," advances a distinction between productive labor, that is, the work of creating and producing goods and ideas, and reproductive labor: childbirth, child rearing, and also the kinds of housework, cleaning, and secretarial work that are the material sine qua non of productive labor.[28] "Automation" is arguably the central term for both Solanas's "SCUM Manifesto" from 1967 and Shulamith Firestone's book-length work *The Dialectic of Sex*. For Solanas, the ideal of total revolution comprises four steps: "overthrow the government, eliminate the money system, institute complete automation and destroy the male sex."[29] For Firestone, too, technology plays a key role. "Eventually cybernation could automate out almost all domestic chores," she writes, and, infamously, Firestone advocates for the automation of birth itself through "(at least the option of) artificial reproduction."[30] She predicts, then, a fundamental separation of reproductive labor from the body, resulting in a "qualitative change in humanity's basic relationships to both its production and its reproduction."[31] That *posthuman* image is catalyzed by scientific and technological horizons of possibility, and ones that were entirely new in the mid-twentieth century: the advance of machines sophisticated enough to transform American life socially and politically. Where, by contrast, Friedan's image of the housewife as robot had been a way to imagine that life as subhuman, Firestone and Solanas both reverse that image so that robots allow for the elimination of devalued women's work. For both of these thinkers, the advance of machines allows us to reimagine the terrain of the human: enjoyment, creativity, collectivity, and affective attachment are newly highlighted as the essentially human qualities in their writings.

But this line of thinking becomes more complicated when we realize that a major component of reproductive labor involves emotion, as care and hospitality work of several kinds that are often called affective labor. The sociologist Arlie Russell Hochschild examines the value of human attachment in her study of these newly dominant forms of affective labor in the postindustrial economy, *The Managed Heart*.[32] Hochschild interviewed flight attendants, whose jobs consisted primarily in creating atmospheres of warmth and comfort, to investigate that increasingly prevalent dimension of women's work in the postindustrial economy. Hochschild emphasizes several times the point that when these workers feel exhausted and no longer feel able to turn on their charm, that they are said to "go into robot" mode.[33] She even, as Alan Liu also notices, concludes the introductory chapter of her book with a speculation that robot jokes might owe their popularity in the early 1980s to this facet of contemporary U.S. life.[34] Hochschild points to a growing segment of the U.S. workforce in which women feel like machines because their job is to *perform* emotions. That performance underscores feelings of "phoniness" and "surface acting" in these occupations and seems to undercut the possibility of leaving undesirable jobs for machines, as Solanas and Firestone had advocated.[35] (Firestone, though, for her part, seems to anticipate Hochschild's concerns where she calls for a *"smile boycott,"* in order for women not to feel as though they must set others at ease in every social interaction.)[36] Curiously, then, at a moment when computers became able to perform logical and computational tasks, and machines could perform a wide variety of physical tasks, the capacity for emotion could be seen as the last sole province of the human and the humane. Consequently, the emptying-out of emotion in emotional labor became an even more apparently insidious form of human automatism as a source of alienation.

Philip K. Dick had influentially interrogated the idea that emotion and the capacity for empathy might be the qualities that distinguish humans from machines in *Do Androids Dream of Electric Sheep?,*[37] a novel built around tests of skepticism between the real and the artificial in people and robots. And in its interest in analyzing the circuits of emotion within and between individuals, Dick's novel also bears out with particular thoroughness Hayles's assertion in *How We Became Posthuman* that the cybernetic concept of the "feedback loop" allows us to see the "boundaries of the autonomous subject" differently.[38] By imagining emotion itself as a set of feedback loops, Dick's novel considers the human brain and body as information processing systems, much like Wiener and von Neumann had done. He then places pressure on the similarities between the computer and the brain when viewed in this mode. The novel's Voigt–Kampff test has been designed to distinguish humans from androids by measuring their physiological reactions, much like a lie detector test measures biometric feedback.[39] When Rick Deckard is testing the opera singer Luba Luft, looking for the "flattening of affect" that characterizes androids (and schizophrenics), he asks her questions that reflect the fictional society's absolute veneration of animals and pets.[40] He becomes frustrated with her evasive answers, despite the fact that he is looking for the disjunctions between speech and physiological readings that would mark Luba as an android. After a long string of evasions, Deckard demands, "I want your social, moral, emotional reaction," and the tension between those qualifiers reveals another possible source of disjunction, a societal dimension to feeling that is distinct from the physiological and the conscious experience of emotion.[41] The novel identifies at least three registers of emotion to which we might give ersatz names: the physiological register of affect, such as sweating, blood pressure, pupil dilation, tears, or a knot in the stomach; the cognition and naming of physiological effects as emotion, which can in turn amplify the corresponding physiological affects ("I'm very sad now"); and the socially mandated or acceptable public "feelings" that we are expected to express (as when at a funeral we might try to suppress, say, laughter or irritability).[42] Between "affect," "emotion," and "feeling" (my terms for bodily, cognitive, and social registers of emotion), no single register is in total control, but rather the totality of what we think of as emotion is produced

Sean Young as Rachel Rosen in Ridley Scott's *Blade Runner* (1982). Here she is undergoing the Voigt–Kampff test, which measures her pupil size and other vital signs, during a conversation that is meant to, but does not, elicit emotional responses from her. In "Rachel" (2009), Larissa Lai will reinterpret Rosen's features and demeanor as explicitly Asiatic.

in the feedback between these registers.[43] That tension between registers of emotion also drives the novel's opening scene, in which Deckard
and his wife manipulate a "mood organ." *Do Androids Dream* overall
defamiliarizes the idea of emotion thoroughly enough to make it seem
like a particularly strange criterion for judging what counts as *human*.

Nonetheless, the skeptical observation of emotion as a performance or
a feedback loop animated many automaton narratives in science fiction,
including those of the not-quite-human robot, such as Isaac Asimov's
robot stories and novels, particularly "Evidence" and *Bicentennial Man*,[44]
and the subplots of *Star Trek: The Next Generation* (1987–94), involving
the android character Data. Emotional connection, a sense of humor,
creativity, or the capacity for pain have emerged as the criteria for humanness in a wide variety of fictions about the boundary between the human
and the machine. Such fictions—most recently Spike Jonze's *Her,* Caradog James's *The Machine,* Gabe Ibáñez's *Autómata,* and Alex Garland's
Ex Machina[45]—examine what machines can and cannot do, as well as
the phenomenological dimensions of human interactions with machines
and other humans. (Indeed, all four of those films highlight a fourth
register of emotion that Dick's novel also explores, that of the interpersonal feedback loops through which affective attachments take place.)

Because these automaton narratives are often structured by the tension between the *surface* of perception and speech and a supposed *depth*
of emotion, humanity, or authenticity, a philosophical fable in the work
of Stanley Cavell is useful in making a final point about emotion and
behavior. Cavell uses a scenario similar to that of Dick's *Do Androids
Dream* or Asimov's "Evidence," in which an observer has to determine
whether a person presented to him is an ordinary person or a "perfected
automaton."[46] In this live-action version of Alan Turing's famous imitation game, Cavell posits that the observer might want to see *into* the body,
a literal version of the depth that is often metaphorically implied. In
Cavell's scenario, the observer opens the body in the anticipation of seeing a ticking clock rather than a heart, but he finds only organs. But the
mystery doesn't end there, as the organs might also be "superficial fak-
ery." In this clever image, Cavell suggests that when we look for "depth,"
what we actually find is another surface. (Dick's *Do Androids Dream*

reaches a similar conclusion with regard to emotion: it turns out that the search for the depth of emotion also reveals only the surfaces of interpersonal and intrapersonal circuits of information.) Cavell's final point in his fable—which devolves to the point where the observer wonders whether the supposed automaton is in pain (depth) or is merely "emitting pain-behavior" (surface)—is that the skeptical search for "human" depth must always fail, leading to a distinction between two ways of seeing others. Cavell suggests that we can only presume that depth by acknowledging others as fully human, because the search for certain knowledge of another's humanness cannot end satisfactorily. Cavell's humanist ethics is one that usefully extends to posthuman beings, and Sherryl Vint writes along similar lines that this kind of acknowledgment is essential for developing community as human bodies become increasingly changed by technology. She writes, "It is imperative that we develop an ethically responsible model of embodied posthuman subjectivity which enlarges rather than decreases the range of bodies that matter."[47] Vint's emphasis in posthumanist scholarship is closely aligned with my own, in that I view the consequences of specific technological modification to the body, mind, or sensorium as relatively minor compared with the consequences of inclusion or exclusion from communities and social networks. The human automaton functions in Cavell's fable and elsewhere as a limit case for the decision to include or exclude.

In that light, we can read many science fiction narratives about humans and machines as fables of acknowledgment that resonate with questions of political acknowledgment. The Voigt–Kampff-style question of whether another person is fully human thus also maps, historically, onto geopolitical questions of who is counted as free or unfree, human or subhuman. For instance, Dick's protagonist Deckard is imagined as unfeeling in a way that is similar to the totalitarian bureaucrat that Hannah Arendt describes in *The Banality of Evil*: "something merciless that carried a printed list and a gun, that moved machine-like through the flat, bureaucratic job of killing. A thing without emotions, or even a face; a thing that if killed got replaced immediately by another resembling it."[48] Indeed, one function of the androids in the novel is to make room for this version of a human automaton indistinguishable from them. Dick's

1972 lecture "The Android and the Human" reinforces this connection, and as with Arendt, the opposite of Eichmann-esque and machinelike evil looks something like civil disobedience: "If, as it seems, we are in the process of becoming a totalitarian society . . . the ethics most important for the survival of the true, human individual would be: cheat, lie, evade, fake it, be elsewhere, forge documents, build improved electronic gadgets in your garage that'll outwit the gadgets used by the authorities."[49] Thus Dick presents another case where antitotalitarian views and New Left progressivism seem to converge, and in a way that tracks the themes and imagery across several works in the first half of his career, including "Second Variety," *The Simulacra,* and *The Three Stigmata of Palmer Eldritch.*[50] Dick's philosophical questions about the "true, human individual" might thus be seen as borrowed from the 1940s and 1950s "crisis of man" discourse that Mark Greif has shown to be a reaction against the totalitarian threat.[51] For Dick, it's totalitarianism proper and "the Man" of the U.S. establishment, hand in hand with technologies based in cybernetics that serve as the bases for that threat.

Before taking a final look at the geopolitical dimensions of posthumanism, I should briefly note an additional way that robots have been used in descriptions of selves and others, the racial association between Asians and machines that Greta A. Niu and others have called "techno-Orientalism." While not a dominant thread in representations of freedom and unfreedom (beyond its instantiations in brainwashing discourse), techno-Orientalism constitutes a consistent racial theme in science fiction that involves Asians. Techno-Orientalism names a form of racialized automatism that revives clichés about Asian Americans, from the machinelike Chinamen of the "Yellow Peril" years of the early twentieth century to the supposed hypnotic powers of Fu Manchu and, later, the Chinese "brainwashers."[52] Niu and the other editors of a recent volume on techno-Orientalism define it as "the phenomenon of imagining Asia and Asians in hypo- or hypertechnologial terms in cultural productions and political discourse."[53] So imagined, such discourse can take on many valences, such as the Samurai-esque mysticism in the Star Wars franchise or the stock yellow-face evil masterminds of *The Manchurian Candidate* or *Iron Man 3* (Shane Black, 2013).[54] From the 1980s onward, this xenophobia was renewed with regard to Asian Americans' supposed facility with computers and technology and with the so-called Tiger economies of East Asia. As the editors of *Techno-Orientalism* point out, the pattern of mainstream representation of Asian factory workers, for instance, often "constructs Asians as mere simulacra and maintains a prevailing sense of the inhumanity of Asian labor—the very antithesis of Western liberal humanism."[55] When framed in these terms, the self- and-other dynamic of human–automaton relationships can be clearly seen in this kind of representation as well.

The 1982 film adaptation of *Do Androids Dream,* Ridley Scott's *Blade Runner,* provides a glimpse into this transvaluation of the human and the machine, as it expresses a 1980s version of the discourse of techno-Orientalism. In a well-known reading of *Blade Runner,* Lisa Lowe de-scribes how the film's vision of dystopia is that of an Asian takeover of San Francisco, and the dystopian Asian takeover seems to mirror that of the main plot's rebel colony androids.[56] Larissa Lai's pair of fictional and poetic responses to *Blade Runner,* both titled "Rachel," imagine that the android Rachel is coded as explicitly Asian American: reserved, seemingly emotionless, enigmatic, piano playing, and innocent looking, played opposite Deckard's expressive and pragmatic gumshoe borrowed from 1940s American film noir. If Dick's Deckard is revealed to be an Eichmann-like organization man, Scott's Deckard appears—before we recognize that he, too, is a replicant—to be a human, American holdout against takeovers by both the Asians and the androids. As Tsarina Prater and Catherine Fung have suggested, Lai's fiction and poem play with the idea of giving voice to a largely silenced figure in the film, though it is the larger trend of techno-Orientalism that makes this pattern of *Blade Runner*'s references legible for Lai's work.[57] And again, here, the self and the other are based around not just the human and the machine but the values of transparently legible emotions and the appearance of freedom (despite the fact that the film's ending destabilizes that freedom and self-transparency). Lai, Karen Tei Yamashita, Greg Pak, and others have writ-ten back against techno-Orientalism in texts that highlight those stereo-types.[58] In Pak's short film *Machine Love* (2003), for instance, scenes of robotic encounter imbue Asimov's and Cavell's scenarios with questions about cross-cultural and interracial understanding. A typical office hires its first robot, Archie, an Asian American android who is much more industrious than his human peers. They are fascinated by him, asking ignorant and offensive questions, and they ultimately mistreat him out of jealousy. The robot Archie seems able to feel this abuse, and he seeks a cure in the companionship of a female robot from another office. Such anti-techno-Orientalist cultural production appoints itself the critical function of revealing such stereotypes and, as in the resolution to *Machine Love,* forming community around this resistance.

Between hypertechnological stereotypes about Asians and those of the mystical, premodern Asia, contradictions and even internal inconsistencies can abound within texts that participate in techno-Orientalism. The hypertechnological subhuman Asian, as in discourse about factory workers, is often figured in terms of the automaton such that both technology itself and the Asians associated with it represent a threat to American humanity. Let's consider how and to what extent science fictional automatism moves beyond techno-Orientalism in *Snow Crash,* a novel that features human automatons while simultaneously embracing both an inclusive multiculturalism and the pro-technological discourses of posthumanism.

Liberal and Fundamentalist Posthumanisms

Although the term *posthuman* has had many meanings over the past three decades, it has been consistently used to imagine the ways that both new technologies and new orientations of thought might precipitate new ways of understanding humanness, agency, our bodies, and our minds. Cary Wolfe's *What Is Posthumanism?,* a book that indicates a range of directions that posthumanist scholarship might take, usefully distinguishes posthumanism from transhumanism, or "the enhancement of human intellectual, physical, and emotional capacities" which "derives directly from ideals of human perfectibility, rationality, and agency inherited from Renaissance humanism and the Enlightenment."[59] For Wolfe, instead, being "posthumanist" is an orientation of thought that "opposes the fantasies of disembodiment and autonomy, inherited from humanism itself."[60] Like Haraway, Wolfe is committed to the strains

of poststructuralist thought that critique and decenter this Enlightenment humanism from various directions. In Wolfe's own work, such a
decentering of the human takes place through animal studies, whereas
in others, it might take place through newly vibrant understandings of
gender or of humanity's place in relation to planetary ecology.[61]

Haraway's cyborg, a primary touchstone of posthumanist scholarship, advances an orientation of thought about feminist possibility that
"skips the step of original unity," such as a vision of femininity identified
with nature, in favor of "partiality, irony, intimacy, and perversity."[62] Haraway imagines feminist science fiction writers as "theorists for cyborgs,"
who explore "what it means to be embodied in high-tech worlds."[63] Haraway holds up as examples Anne McCaffrey's *The Ship Who Sang*,[64] as
an early fiction of a body–machine hybrid where the body does not "end
at the skin," and Vonda McEntyre's *Superluminal*,[65] as a story where a
woman has her body modified to pilot a spaceship.[66] Haraway does not
imagine, though, that bodily modification must necessarily lead to a
revolutionary subjectivity. The 1970s television show *The Bionic Woman*
can serve as a simple counterexample, wherein the heroine Jaime sur-
vives a parachuting accident and is given bodily modifications and begins
working as a spy.[67] While Jaime's body is enhanced through modifications, her mind and outlook are more or less unchanged. The show's
emplotment of technology highlights a common theme in science fiction:
the tension between a technology's utopian potential, embodied in Jaime,
and its dystopian double, figured in *The Bionic Woman* by Jaime's recurring enemies, "Fembots," who enact the unfreedom and evil of robotic
subjectivity and whose face plates are frequently removed to reveal the
void of electronic circuitry.

Wolfe's distinction between transhumanism and posthumanism
proves useful, then, in distinguishing between the physical modifications
to the body that Haraway's manifesto had used as examples and the conceptual orientations toward the category of the "human" toward which
Haraway's examples had pointed. In a fruitful path beyond this distinction between modified bodies and modified orientations of thought,
Hayles's work on the posthuman and on "technogenesis" asserts that our
perceptual faculties have evolved more subtly alongside shifts in media

technology. Whereas fictional narratives about human augmentation by computer often foreground the wholesale modification of the mind in dystopian scenarios of human programming, Hayles's posthuman and technogenetic human simply have slightly different emphases in their perceptions of the world around them, such as the desire for disembodiment that the many manifestations and imaginations of cyberspace or virtual space entail.[68] Like Wolfe and Haraway, Hayles maintains an interest in thinking beyond the confines of the conventional humanist subject and even laments the conventionality of some of the science fictional works she takes up in *How We Became Posthuman*. She notes, with regard to four of the works she reads there, that "many attributes of the liberal humanist subject, especially the attribute of agency, continue to be valued in the face of the posthuman."[69] Hayles finds in *Snow Crash* a contrast between the human and posthuman that looks much more stable than in Philip K. Dick's work, where "automata . . . body forth a version of the posthuman that stands in horrific contrast to the free will, creativity, and individuality that for Stephenson remains the essence of the human."[70] To see this as backward looking is to raise a historical conundrum that Hayles notices with regard to the novel's ancient Sumerian backstory: the novel claims, she writes, that "the best way to counteract the negative effects of the posthuman . . . is to acknowledge that we have always been posthuman."[71] It seems that instead of thinking in terms of a human past and a posthuman future, we might do just as well to trace how technology has been put to use in representations of human selves (and sometimes technologically transhuman or superhuman selves) and posthuman others (who are, by dint of their machinelike nature, subhuman).

Snow Crash can be fruitfully revisited as a commentary on Haraway's "cyborg" and Hayles's posthuman, because the novel's characters map onto a wide variety of points on the spectrum between man and machine, as a sort of "character system" that Fredric Jameson has described in Philip K. Dick's novels.[72] In rough order from man to machine, Stephenson's typology of the cyborg in the novel includes (1) Hiro, who leads a second life in cyberspace and whose ability to read binary code makes him susceptible to the snow crash virus; (2) YT, a "gearhead" who harpoons cars on her skateboard-like device and wears many gadgets, including an antirape device; (3) YT's mother, a federal employee who is forced to take regular polygraph tests and whose movements are entirely monitored by a computer surveillance system; (4) Lagos, a man who wears surveillance gear to collect intelligence all around him; (5) Ng, a paralyzed man who has been wired into a van that functions as a prosthesis and who enjoys a full body in cyberspace; (6) Rat Thing, a guard-dog robot with some biological animal parts who lives half in cyberspace and half in the real world; (7) the Librarian, an AI engine with a cyberspace body and personality who serves as a voice-controlled interface to a large database; and (8) the Raft people, drones who have antennae drilled into the bases of their skulls, who babble nonsense and are controlled by the tycoon L. Bob Rife.[73] For all these figures, their lifestyles, and usually their minds, are hooked into machines of one sort or another, but the modalities of subjecthood are distinctive for each.

Ng, whose cyberspace avatar includes virtual masseuses who massage his cramp-prone back, is perhaps the most universally laudable kind of cyborg in Haraway's sense: in an earlier era, his movement would have been severely limited, so the notion of his virtual mobility within a "normal" cyberspace body, and an innovative and differently abled van-body hybrid that simply enables him without reference to traditional standards of bodily normality. (He might also conceivably be categorized as "transhuman," but he resembles Haraway's examples closely in resisting normative embodiment and in achieving a new outlook through his new embodiment.) A reader of Haraway could conceivably point at Ng as among the best exemplars of the liberatory cyborg. But the reenabling of the body is far from the only kind of human–animal–machine hybrid here. On the other side lie, for instance, the Raft people, who behave a bit like video game enemies, and the Librarian, who functions within the fiction as a conduit for exposition about the Snow Crash virus's Sumerian backstory. Considering the poles of a natural body and a mechanical body, on one hand, and a free mind and a programmed mind, on the other, we arrive at the system depicted on the next page. Although this system leaves out the important dimensions of cyberspace and surveillance, it can help to indicate that the axis of sympathetic identification with characters falls between free minds versus programmed minds, while the bodily compositions of the characters are relatively unimportant.

In this scenario where bodies matter less and less, *Snow Crash* continually emphasizes protocols of openness and acknowledgment between characters of different kinds. When the skater YT encounters the cyborg Rat Thing security dog, she is uncertain whether it should be seen as a subject worthy of sympathy or as a lifeless object. When it malfunctions in the course of a high-speed chase, the robot begins to overheat. YT sees it struggling on its back, its mechanical guts exposed: she can

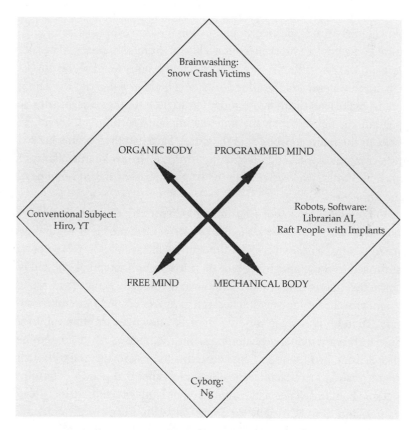

The possible combinations of cyborg and conventional bodies and minds as they appear within the character system of Neal Stephenson's *Snow Crash* (1992).

either dissect it and sell the information, or she can help it, risking getting burned by the steaming-hot specimen. In a single sentence, she
embodies two ways of seeing, of knowing and acknowledging: "This is
intel; poor thing is burning itself alive."[74] In this variation on Cavell's
scenario of the automaton in pain, YT decides to help it by returning it
to its air-conditioned home base, and, in typical action-film fashion, the
dog appreciates the gesture and comes to YT's aid later in the novel. This
is by no means an extraordinary sequence of events but a good example
of the kind of interactivity with nonhumans that takes on the character
of community, and another instance in which Vint's ethical orientation
toward the question of "posthuman" subjects proves useful. Stephenson's
protagonist Hiro is, despite all his surfing through cyberspace and technical know-how, a fairly conventional liberal–humanist subject. The subjects on the sympathetic—conventional and cyborg—end of the novel's
character system, and the ethical concerns highlighted in the Rat-Thing
scene, make it all the more important to account for the nonsympathetic
posthuman figures: first, the antenna-implant-sporting cult members
who, in the main conspiracy of the novel, have had their brains hacked
by a computer virus, and second, the mythical, dronelike inhabitants of
ancient Sumer, who suffer from the language-based ancient versions of
the same virus.

In the novel, we do not see from the perspectives of these characters,
nor do we sympathize with them at all. In fact, we see them as subordi-
nated groups, as members of a fundamentalist sect, and we are not so
surprised to see them stripped of their free will. If anything, the radio-
controlled goons that populate the novel's climactic action fight scene
might provoke the kind of mild horror typically associated with contemporary action films or video games. American audiences are often comfortable with demonizing a posthuman cult member or mythical zombie
people from the Middle East, and the difficulty of sympathizing with such
characters marks out a dialectic of self and other that plays a determining role both in science fiction and in other kinds of genre fiction, video
games, and films. Crucially for *Snow Crash*, the image of the antenna-
implanted Raft people borrows heavily from cybernetics and information theory. Claude Shannon's information theory indirectly provided
a useful and widespread template for imagining the spread of ideas in

contemporary culture. The notion, implicit in Burroughs's work, that an idea might be like a virus, that it might program and take over its human host, found its way into philosophical discourse in 1976 through the work of Richard Dawkins, in *The Selfish Gene*. Dawkins uses the metaphor of the gene as a program throughout the text, primarily as a grammatical means of making genes coherent wholes, for example, "a program may take the form of the following instructions to the survival machine," that is, the organism, for whom the instructions are instincts.[75] Whereas Orwell's totalitarian language had needed the apparatus of a totalitarian censorship apparatus to limit ideas to orthodoxy, this viral language of Burroughs's and Dawkins's imagining virtually spreads on its own. Throughout *The Selfish Gene*, Dawkins uses the modern computer program, such as a chess-playing program, as an analogy for programs or ideas replicating themselves in a way that might be imagined analogously to genes. It is in these descriptions that Dawkins coins the word *meme*, and the first example he gives for this metaphor for the programmability of the mind is anything but innocent:

> The survival value of the god meme in the meme pool results from its great psychological appeal. It provides a superficially plausible answer to deep and troubling questions about existence. . . . The idea of God is copied . . . readily by successive generations of individual brains. God exists, if only in the form of a meme with high survival value, or infective power, in the environment provided by human culture.[76]

In this particularly striking imagery, the "environment provided by human culture" more readily resembles the "culture" of a petri dish, through which a microbe or crystalline structure might grow, unimpeded. In the first sentence quoted, the minds of humans, taken collectively, constitute this "meme pool." The "individual brains" in which the idea of "god" takes hold are mere hosts. Such a notion translates, too, the cybernetic approach to information and coding (which had been a key factor in the decoding of DNA) as concepts applicable to any transfer of information whatsoever. When the human "host" becomes a relay station for information in this passage, Dawkins treats the human subject as more of an intermediary—that is, as a passive conduit for information—than as a mediator of the information that it propagates.[77] This sort of *flat* human subject, for Dawkins, allows us to imagine the human as an intermediary for an idea in the same way that we would load the same program onto several different computers and expect them to work in near-identical ways. Like Ellison's technocrats and Burroughs's Nova criminals, Dawkins makes the choice to see the human subject as a passive intermediary for active information that is spread like the replication of genes, programs, or viruses.

This "meme" orientation toward the spread of a virus that's also a fundamentalist religion absolutely defines *Snow Crash*'s approach to the dystopian form of posthuman subject. One character asks another, "This Snow Crash thing—is it a virus, a drug, or a religion?" only to be answered with another question: "What's the difference?"[78] A passage late in the text describes some of these Raft people, the group who have had radio antennae surgically implanted into their brains, much as the intermediaries, which Dawkins might describe as transmitters of a meme:

> A single hair-thin wire emerges from the base of the antenna and penetrates the skull. It passes straight through to the brainstem and then branches and rebranches into a network of invisibly tiny wires embedded in the brain tissue. Coiled around the base of the tree. Which explains why this guy continues to pump out a steady stream of raft babble even when his brain is missing: It looks like L. Bob Rife has figured out a way to make electrical contact with the part of the brain where [the Sumerian god of text] Asherah lives. These words aren't originating here. It's a pentecostal radio broadcast coming through on his antenna.[79]

The memelike notion of words that "aren't originating here" in the individual subject allows for the same sort of uncanny effects as automatons do for Ernst Jentsch, when characters attempt to speak to these apparently conscious beings. Although they initially appear to be interacting as humans, these Raft people behave more like objects than some of the computers in the text do. In the positive and negative posthumanism that the novel presents, then, we see what ultimately come to utopian and dystopian poles of technologically enhanced humanity: either one becomes a cyborg, and technological modifications to the body or the senses enhance the range of freedoms one can enjoy, or, in the fanciful dystopian version, technology reaches a new horizon in which individuals' freedoms are wholly destroyed. These utopian and dystopian poles are presented easily within the bounds of genre in *Snow Crash*, where enemies with no agency prop up the sublime agency of the typical action film, in which the good guys manage to save the world.

Throughout the broad circulation of computer-inflected human automaton figures in postwar U.S. science fiction and culture, cultural producers' decisions about the distinguishing features of humans as opposed to machines often tell us much more about the authors' values than about the difference between humans and machines. Whether that difference comes in the form of expressing emotions, understanding humor, creative knowledge work, or engaging in civil disobedience, computation provides a technological substrate for representing human automaton figures who lack these features. Crucially, then, the shape of *Snow Crash*'s world also bears the traces of U.S. post–World War II history, as it continues a dialectic of free American selves and unfree others that the final two chapters of *Human Programming* will chart through the last several decades. Read through its historical moment of 1992, *Snow Crash* functions as a savvy analysis of a post–Cold War moment when a dominant opposition between Western cosmopolitan multiculturalism and Orientalized, closed fundamentalism emerged at the so-called end of history. Because it merges the freedoms of new technology with the global circulation and exchange of culture, *Snow Crash* also presents a compelling and largely optimistic vision of multiculturalism as the cultural expression of that moment in technological and geopolitical history. Richard Rorty and Walter Benn Michaels have both offered readings of the novel that take the multicultural theme as central. Rorty positions the novel as largely dystopian, characterizing the Raft as a "vast international slum" full of "Asians" whom Americans view with "relief at being safer and better-fed than those on the Raft."[80] But with several prominent and relatable Asian American and Asian characters (including the main protagonist Hiro), the novel is less biologically xenophobic or techno-Orientalist than this description would suggest. Michaels for his part sees in the Raft people's virus ("a virus that infects you does so not because of what it means but because of what it is") a demonstration of a multicultural logic of meaning: "the world in which everything—from bitmaps to blood—can be understood as a 'form of speech' is also a world in which nothing actually is understood, a world in which what a speech act does is disconnected from what it means."[81] Michaels downplays or ignores the novel's strong contrast between fundamentalism (to which the virus and blood-oriented characters like Raven are tied) and an elective, consumerist model of multiculturalism, where cultural mixture is a matter of the free choices individuals make. The novel imagines many different forms of hybridity and cultural mixing: Hiro Protagonist is half African American, half Japanese, and he listens to Russian "fuzz-grunge" and a Japanese rapper named "Sushi K," and different "burbclaves" are set up so that individuals can choose which culture they want to be a part of. Hiro, along with Ng and YT the young gearhead girl, are also like Haraway's cyborgs in that they embody a progressive and hybrid view of culture, far removed from racial or gender essentialism. The novel's villains are Raven (an aggrieved Aleutian terrorist), L. Bob Rife (a corporate magnate and evangelical church leader), and the Raft people—all of whom are associated with intolerance and fundamentalism. It is here that we see, going into the post–Cold War context, the adaptability of the human automaton in the range of both technologies that it animates within fictional texts and politically charged resonances it might hold.[82] Although Stephenson's novel does draw on the tradition of techno-Orientalism in its Sumerian backstory, it takes care to include Asian and Asian American characters who are part of an American multicultural community. More in keeping with American Cold War texts, automatism is ascribed to the programmed, curiously unempathetic minds of those who refuse democracy. This time, it is the automatism of fundamentalists. To further explore that conflict at the end of the Cold War between the values of a capitalist, democratic, multicultural cosmopolitanism and an anticapitalist, religious fundamentalism, though, we will need to encounter the scientific and public narratives that developed, from the 1970s onward, the metaphor that Stephenson had literalized in *Snow Crash*: cult programming.

CHAPTER FOUR

Cult Programming

Extremism, Narrative, and the
Social Science of Cults

The narrative conventions surrounding the cult are simple but effective. In the "True Believer" episode of Joss Whedon's science fiction series *Dollhouse* (2009–10), for instance, Echo, the programmable agent, is sent to help law enforcement officials root out a charlatan cult leader who has committed human rights abuses within his compound.[1] In the opening scene, all the cult members enter a country convenience store singing in unison, with beatific but vacant smiles on their faces. After the transaction has taken place, the clerk notices a receipt that has been passed to him with the word "Help!" scrawled on the back. The eerie appearance of calm and joy, which had angered and unsettled the other customers in the store, is unmasked as a facade. Special agent Echo infiltrates the compound in disguise to find out what is really happening in this closed space. She finds the cult members reluctant to talk about their problems with an outsider, and she is faced with several difficult questions: are they terrified captives, so horrified by their surroundings that they are afraid to speak, or have they been manipulated into enjoying conditions that are objectively detrimental to them? And how, precisely, are those two states different? And, moreover, if the cult members' belief is genuine, what business do law enforcement agencies have interfering with it at all? Once these questions are raised, the charlatan leader is fully exposed, and the compound goes up in flames. Many of the characteristics of this particular cult in *Dollhouse* echo those of the Branch Davidian compound in Waco, Texas, in which the leader, David Koresh, faced allegations of sexual misconduct and child abuse. In 1993, a shoot-out with federal agents ended the lives of the leader and many members of Koresh's group, a radical faction of the Seventh Day Adventists.[2] Although a span of eighteen years separated the Waco incident from the filming of *Dollhouse,* the space of the rural cult compound clearly resonated strongly enough to mirror it closely in television.

As a science fiction series, *Dollhouse* imagines the cult as a space of social programming and automatism alongside the imaginary human automatism of Echo's computer-programmable mind, but the conventions of cult narratives can also work without this science fictional premise.[3] Many other texts about cults—they are a staple of contemporary crime dramas on television, in particular—raise the same set of questions about extraordinary beliefs, closed forms of society, and the ethical consequences of the programming or manipulation of group members, often with details borrowed from the stories of real cults. Several factors make the cult episode a satisfying narrative formula: the charlatan cult leader makes an excellent villain, just as the entrapped members are convincing victims; social and political issues such as domestic and sexual abuse, religious freedom, and the limits of tolerance are ready to hand; and the exposure of the cult as fraud is always thrilling. The larger questions that cults raise are as accessible as they are insoluble.

A term whose usage has been expanded to comprise both political and religious communities whose beliefs are extreme, the word *cult* has been used to describe groups as diverse as the Symbionese Liberation Army, the Unification Church, the Jim Jones People's Temple, the Branch Davidians, and the Heaven's Gate sect. Cults are often imagined in conjunction with techniques like hypnosis, torture, and psychological manipulation, but representations of them sometimes focus wholly on the power of the leader's charisma and persuasiveness. (The recent films *The Sound of My Voice* [Zal Batmanglij, 2011], *The Master* [Paul Thomas Anderson, 2012], and *Kumaré* [Vikram Gandhi, 2012], for instance, are all most invested in the complementary psychological needs of the cult members and the leaders.) The cult is both a culturally exotic space and a contested analytic category within sociology and religious studies (where cults are more often referred to as new religious movements [NRMs]). We might either imagine the cult as an extreme form of social life, wherein the quirks of ordinary culture are suspended or amplified, or as a principled refusal of that ordinary culture. The words *programming* and *brainwashing* are both part of the cult image, which is often depicted as a space outside a contemporary secular culture of diversity, tolerance, and hybridity.

The widespread representation of cults in that mode dates from the 1970s, when a flurry of high-profile NRMs formed, flourished, and made headlines in the United States. In this chapter, I describe the development and the particularly wide cultural circulation of what I call the *topos* of the cult. The idea of the cult has developed, as with other supposed sites of human automatism in U.S. culture, through a set of exchanges between popular culture, scientific research and writing, literary and cinematic forms, and news media. Scientific research in psychology and sociology formed the criteria for the *topos* of the cult, which describes a form of society that is closed off from communication with the outside, headed by a charismatic leader, and organized through techniques of coercive persuasion, often including specialized jargon that binds members together. I use the word *topos* to emphasize that the dominant idea of the cult entails both a set of narrative conventions and a shape or *form* of social organization, one whose continuities with the closure and control of totalitarian society have been continually emphasized by experts, reporters, literary and popular fiction writers, and ordinary citizens alike. These representations of cults thus constituted a new development for the human automaton figure in U.S. culture, where specialized spaces and conditions might make possible the kinds of unfreedom Americans had until then primarily imagined in relation to totalitarian states, POW camps, total institutions, and science fictional fantasies of human programmability. Although depictions of cults sometimes represented and interrogated the automatism of cult members directly, as the trial of Patty Hearst would do extensively, the social dynamics of that *topos* of the cult became a focal point for imagining the contours of mental unfreedom on U.S. soil from the 1970s to the 1990s.

The *Topos* of the Cult

The contemporary and primarily derogatory use of the word *cult* dates from the mid-twentieth century. Earlier, the still-current usage of the term within religious studies and art history, as a group with a particular object of devotion, applied more broadly: the cult of the Virgin Mary or the cult of Apollo or the cult of personality. Mark Twain, for instance, wrote an impassioned and highly critical book about Christian Science in 1907 that uses the term *cult* only in passing and never with a derogatory connotation.[4] As late as 1963, a book on cults by Anthony A. Hoekema would be titled *The Four Major Cults: Christian Science, Jehovah's Witnesses, Mormonism, Seventh-Day Adventism*.[5] Although *cult* here is not exactly a term of endearment, it refers to Christian sects outside of the larger and mainstream denominations. By 1973, however, William J. Peterson's *Those Curious New Cults* would reflect the still-current understanding of cults as NRMs, and that book's sections would address Dianetics, the Church of the Final Judgment, Yoga, the Esalen Institute, Soka Gokkai, the Rosicrucians, Hare Krishnas, and the Baha'i.[6] The late 1960s and the 1970s would see more publicity for cults as well as the rise of an extensive young adult and children's literature of cults, much of which warned young people against joining cults and expressed a strong current of parental anxiety about the persuasive power of these groups.[7]

The scientific study of cults in religious studies, sociology, psychology, and psychiatry reveals some of the interpretive tensions that surrounded the groups in the past and today. Behind one terminological difference among sociologists and religious studies scholars—are these groups "cults" or "new religious movements"?—lay a host of assumptions about these groups' natures, and particularly the question of whether they were a social problem to be studied and solved or a form of religious diversity to be understood and protected. The violence that surrounded some groups, most infamously the mass suicide of Jim Jones's People's Temple in 1978, drove many researchers to treat cults as a social problem and to investigate their techniques of "brainwashing."[8] In scholarly debates about whether it ought to be called "brainwashing," however, proponents of seeing these groups as NRMs have noted that anticult researchers have tended only to interview ex-members of these groups, who tend to exaggerate their grievances, and that "cult stereotypes" might prejudice them against any NRM whose practices may be fulfilling and rewarding for members of the group.[9] That particular obstacle to objectivity and certainty extends, too, to journalism and nonfiction. After Lawrence Wright's recent and extensive account of Scientology, *Going Clear: Scientology, Hollywood, and the Prison of Belief,*[10] officials from that group claimed that Wright's evidence and interview pool were skewed and contained many falsehoods.[11] And in *Stories from Jonestown,* Leigh Fondakowski interviews survivors and others connected to the People's Temple who remain uncertain about the psychology of the place and for whom the question of whether it had been "brainwashing" or "true Christianity" remains difficult to puzzle through.[12] That interpretive impasse also emerged in U.S. courtrooms in the 1970s, as I describe it relative to Ted Patrick in the following pages, as a question about the limits of religious freedom.

The pathological status of cults is most frequently defined through criteria that conceive of the *topos* of the cult as a peculiar, cordoned-off social space. The psychologist Margaret Thaler Singer has stated that the cult is identified by three main characteristics: (1) the use of coercive persuasion (i.e., brainwashing or propaganda techniques), (2) the closure of communication with the outside world and the use of specialized language, and (3) charismatic or quasi-divine leadership.[13] These criteria appear in many forms in anticult literature but can be traced back to the work of Robert Jay Lifton, who had also been an expert on Cold War "brainwashing." In his well-received book *Thought Reform and the Psychology of Totalism: A Study of "Brainwashing" in China,* the Harvard-based psychiatrist had taken on Edward Hunter's project of describing "brainwashing" in a more scientific direction, and with the word "totalism," he also borrows from the psychologist Erik Erikson a way to describe totalitarian psychology outside of totalitarian states.[14] Lifton emphasizes there the spatial criteria that Singer later echoes, as well as specifically linguistic criteria for cult language, a schema that shares common threads with Orwell's NewSpeak. Lifton points to "milieu control" (control of communication) and a "demand for purity," a linguistic culture that involves a "cult of confession," an "aura of sacredness," and the "constriction" of group language.[15] In addition to its own social formation, then, the cult has its own linguistic or even literary style. In both its spatial closure and its specialized language, the cult can be said to reproduce the strategies of totalitarianism in miniature. Lifton offers as the alternative to totalism an explicit reference to Lionel Trilling's liberal imagination, an "essential imagination of variousness and possibility" holding the "awareness of complexity and difficulty."[16] In Lifton's work, the values that Trilling associated with the novel's encouragement of cosmopolitanism and critical thought are a touchstone of democratic self-definition. By replicating the form of the totalitarian state in miniature, the cult's brand of unfreedom is seen as deriving from its nondemocratic *form* of society rather than from the elements of its religious doctrine. This continuity between different descriptions of the *topos* of the cult has been shared through the cult's public history in the media as well.

The controversial nature of cults became particularly well established in the U.S. news media in 1973, when a series of trials in New York City involving the "father of deprogramming," Ted Patrick, raised questions about the legal rights of NRM members. Patrick began his career deprogramming members of the Children of God, which was, according to the *New York Times,* an "ultra-fundamentalist communal sect that—according to some parents—'brainwa[sh]es' its followers [to] turn over their worldly goods to the organization."[17] In the first of Ted Patrick's high-profile trials, Daniel Voll's parents had enlisted Patrick to retrieve their son, a Yale undergraduate whom they thought had joined the Children of God. (He had in fact joined the New Testament Missionary Fellowship.) At the parents' behest, then, Patrick kidnapped Daniel Voll as part of one of these "deprogramming" operations. Patrick's career of deprogramming would eventually comprise hundreds of kidnappings wherein various regimens of coercive persuasion would be used to bring young adults back out of cults and into the protection of their concerned parents. At Patrick's trial for the kidnapping and assault of Voll, the prosecution brought a witness to describe Patrick's technique. This witness, Arlen Thorpe, had been unsuccessfully deprogrammed from her membership in the Tony and Susan Alamo Christian Foundation in one of Patrick's earlier cases. Thorpe had been held captive in a motel room for more than ten days, during which "up to a dozen people at a time subjected her to a 'constant verbal barrage' for up to 14 hours a day."[18] Although Patrick was acquitted on the Voll case, he was in and out of court throughout the 1970s, and he gained a widespread reputation, particular among parents of cult members, for his work. Andreas Killen has recently revisited Patrick's role in the culture of the 1970s, imagining him as a part of an intensified intergenerational battle between parents and their children during the period. During the "me" decade, Killen argues, films like *The Exorcist* (William Friedkin, 1973) imagined youthful rebellion as the result, not of growth into adulthood and independence, but of malign possession—in this light, Ted Patrick's role can be read as something like that of the exorcist of the film, restoring children to innocence and to their parents.[19]

Patrick's technique to counter the supposed cult brainwashing was for the most part an attempt to create a more intense form of brainwashing, an irony lost on few at the time. In place of the *topos* of the cult in the Children of God, Unification Church, and many other sects— their charismatic leaders, their cutting ties with parents and old friends, and their new systems of belief—Patrick substituted a more intensified, shorter process that fell somewhere between counseling, rehab, and torture. As he describes his first deprogramming in his memoir, Patrick describes how a "Biblical debate" with a young member of the Children of God took place nonstop until, "after two days of talking . . . she suddenly gave in. She snapped, just as if someone had turned on a light inside her."[20] That description of "snapping" holds a great deal in common with what scholars of torture call the "break," the moment when outside reality seems to disappear for the victim. Elaine Scarry's *The Body in Pain*[21] took up the ethics and psychology of torture in terms that again echoed Lifton's, and her study of real-world torture programs emphasized language games, intensified interpersonal relationships, and the closure of social space taken to the limit. In that sense, the torture chamber belongs within the continuum of closed social spaces, like the totalitarian state and the cult, that are widely imagined to be prerequisites for coercive persuasion. Brainwashing and Patrick's deprogramming practice, too, resemble Scarry's account of torture in that they constitute scenes in which individuals are made more malleable by way of a highly orchestrated process of manipulating social space. "In torture," Scarry writes, "the world is reduced to a single room or set of rooms."[22] If the cult resembles a totalitarian state in miniature, then the scene of torture constitutes a theater of even more precise control, and a scene that literary and cinematic texts would take up increasingly in the 1990s and 2000s. Scarry explains, using nonfiction reports on torture techniques from around the world, how techniques similar to those of the brainwashers— sleep deprivation, the infliction of pain, and language games—have been used to create the torture chamber as an exceptional, and closed, space. Regimes of renaming ordinary objects accompany pain in these scenes, and Scarry analyzes the "world-ridding, path-clearing logic" of pain as it is enlisted in a linguistic transformation of the victim's world.[23] Across many regimes, torturers make up nonsensical names for the victim's pain ("the motorola," "the plane ride," "the birthday party") and the scene of torture ("guest rooms," "safe houses," "the parrot's perch"), while using everyday objects as instruments of torture (filing cabinets, refrigerator doors), devices that Scarry claims "annihilate" the victim's "civilization" and, in a phenomenological sense, the victim's "world."[24] This analysis gives rise to an innovative way of considering the relationship between subject and environment, in which surroundings, everyday objects, and language in fact constitute part of the subject that the scene of torture destroys. That phenomenologically expansive understanding of the individual subject can in turn make more sense of the *topos* of the cult as a form of social organization and a set of techniques—such as intensive confession sessions, the introduction of new vocabularies, and the severing of ties with family and friends—designed to facilitate a significant transformation and reduction of the cult member's physical and social environment. As in my discussion in chapter 2, the totally controlled space—either in the total institution or in the totalizing social space—creates the conditions of possibility for forms of consciousness wholly determined by their environment.

In his memoir, Patrick is careful to distinguish his techniques from those of torture and brainwashing, even as he admits that "limiting sleep is a basic element in deprogramming," just as sleep deprivation is a "strategic weapon" for the cults themselves.[25] The difference, he claims, is that he tries to "show the victim that he has been deceived," and, unlike the cults that "prevent the person from thinking," Patrick does all in his power to "start [his subjects] thinking."[26] Patrick first developed this technique, what he calls "fighting fire with fire," after having gone undercover into the Children of God. His own son had had a long encounter with the group, and in his capacity in the California state government, as special representative for community relations in San Diego and Imperial Counties, he had fielded several parental complaints about the group. In summer 1971, Patrick went out in search of the Children of God. Although the group being recruited onto a bus at Mission Beach was predominantly made up of white teenagers, it was heterogeneous enough that Patrick, a forty-one-year-old African American man, did not arouse the recruiters' suspicions. When he arrived at the compound, he found intensive persuasion techniques that all seemed geared toward convincing the members to disclose what possessions they owned and how much money was in their banking accounts. He saw, in the open, what he estimated to be about one hundred thousand dollars' worth of members' worldly possessions that had been donated. The organization's concern with a prospective convert's financial situation was obvious to Patrick as an outside observer, but he recounts the confusion recruits experienced while filling out a financial questionnaire. The recruits were simultaneously "queried directly about your financial status, . . . listening to tape recordings of Bible verses, being exhorted by one member to pray and praise the Lord, and being hugged by another member who tells you he loves you, brother—so it is all very confusing and one does not use his powers of concentration or critical ability in a normal way."[27] The deprogramming operations, as described by Patrick and the media, differ primarily in that they involve less distraction, as Patrick here insists that the group's aim has been to overwhelm, distract, and "confuse" to take financial advantage of the groups. Clearly the financial motivation, which Patrick here claims to have witnessed, renders the groups he infiltrated suspect.

TIME magazine's 1973 story on Patrick's methods gives a sense of the mystique and the cultural resonances of the "deprogramming" operation.[28] The article mentions both older monastic religious orders and Stanley Kubrick's *A Clockwork Orange* to put "deprogramming" in a larger context. Indeed, the cinematic scene of Kubrick's Alex, strapped in the cinema with eyes pinned open, is an evocative scene of human programming that would have resonated with readers, even though Patrick would deny in his later memoir that Kubrick's film inspired his own methods.[29] Rather, for readers of *TIME*, this cinematic scene would have lent an added mystique to Patrick's "deprogramming," a scene whose dynamics are familiar and yet whose particulars always remain mysterious. The mystique of such scenes constitutes a feedback loop between scientific processes, science fictional imaginings, and Patrick's own earnest reenactment and restaging of what he considers to be a scientific scene. Given that Patrick states his mistrust of professional psychiatrists near the end of his memoir, his method for deprogramming is something like a *vernacular* scientific practice, Gramsci's organic intellectual as vigilante.[30]

In 1974, Patrick would be tried again, on charges of false imprisonment. The subject of this deprogramming, however, would be of a different kind from his normal clientele: it would be a member of a radical Marxist group, the National Caucus of Labor Committees (NCLC). Patrick told the *New York Times* of this instance of *political* deprogramming: "These religious fronts and political fronts use the same techniques of destroying a person's free will. . . . They just have different names. . . . She was psychologically kidnapped" and subjected to "on-the-spot hypnosis and brainwashing and mass psychology, same as Hitler and Red China."[31] The young woman's parents would describe the situation in much the same terms as other parents of cult members used: "she was mesmerized and put into a zombie-type situation by that group." The flurry of terminology used even in a single newspaper article about the case points to a fundamental uncertainty about precisely what happened to this woman, identified by the *Times* only as "Miss Roeschman," within the NCLC political group. At the same time, they admit a certainty, on Patrick's part and others', that these processes, spaces, and situations of unfreedom are fundamentally interchangeable: religious and political sects; sects and cults; and children who, in the image of the cinema, have been mesmerized, hypnotized, brainwashed, or zombified. As a crusader against these forms of 1970s American unfreedom, Patrick would become a minor player in the largest political "cult" drama of 1974, 1975, and 1976, when he visited San Simeon, California, to offer his deprogramming services to Mr. and Mrs. William Randolph Hearst.

"Six Weeks to Straighten Me Out": Patty Hearst and Brainwashing

The 1970s mystique of cults and brainwashing met with discourse about radical politics most forcefully in the 1976 trial of Patty Hearst. Hearst, a college-aged newspaper heiress, was kidnapped by the Symbionese Liberation Army (SLA) in 1974, joined the SLA months later, participated in a bank robbery and other crimes, and then was captured and stood trial in 1976. In this high-profile case and widely circulating narrative about coercive persuasion, scientific research on communist brainwashing came into contact with the uncertain situation of post-1960s political radicalism. Hearst has long been described as a transitional figure: in Joan Didion's account, as a last gasp of the culture of the 1960s; in Nancy Isenberg's, as the first "postmodern legal subject" (looking to her nebulous identity and its complex media coverage); and, in a recent book-length treatment from William Graebner, as a figure suspended between 1960s "victim" culture and the Reaganite 1980s culture of "responsibility."[32] Though she was all of these things, her trial also showed the crystallization and failure of forms of expertise around brainwashing and cults, which were beset by many of the same uncertainties about states of unfreedom that had been on display in television, film, and fiction since before the Second World War. Hearst herself, in the tape recordings and video surveillance that captured her voice and movements, would become the most memorable human automaton of the 1970s.

Following the kidnapping, Hearst spent three months blindfolded in a closet, being reeducated, harassed, and sexually abused by SLA members, before she stated in a scripted communiqué that she had chosen to join the SLA and fight for "freedom."[33] In 1976, after Hearst had been apprehended for the Hibernia Bank robbery, during which she had been taped by surveillance cameras, her star defense attorney, F. Lee Bailey, embraced a brainwashing defense. Bailey called several of the psychological experts on Cold War–era brainwashing to the stand as the backbone of this defense strategy, who each in turn established the criteria and diagnoses of the coercive persuasion and traumatic neurosis that would have diminished Hearst's free will and led her to follow the SLA only because of her traumatized state. Among these expert witnesses were the scientists mentioned earlier, including Robert Jay Lifton and Margaret Thaler Singer. Both Lifton and Singer explored in their work from the 1970s to the 1990s the Manson Family, Jim Jones's People's Temple, the Unification Church, Heaven's Gate, and the Japanese group Aum Shinrikyo. Lifton had already done research on cults by the time of the trial, and he mentions that research as relevant professional experience in addition to his research on Korean War POWs, because, he states, "the extremist religious cults, like the Unification Church, Children of God, and a few other fundamentalist religious groups . . . have made use of a process that resembles, in some ways, some of what I have studied around thought reform or coercive persuasion."[34] Additionally, the defense's expert witnesses all dealt in some degree with the Korean War POWs, who had been the first subjects to be called brainwashed. That set of resemblances, between POW brainwashing, cult programming, and Hearst's situation, made up the rationale for the brainwashing defense, and frequent comparisons with POWs and the Manson Family murders were also made throughout the trial.

One of the most curious things about the trial, and a phenomenon new in the post–World War II period with the ubiquity of recording apparatuses, was the proliferation of recordings of Hearst. The court had many tapes of her making revolutionary statements and, most famously, the closed-circuit television surveillance footage of Hearst holding a gun during the Hibernia Bank robbery, which was also submitted as evidence. Head prosecutor James Browning noted during the trial, "Rarely has so much evidence of apparent intent been available to a jury of a defendant having participated in a bank robbery as we have in this case."[35] But his wording—*apparent intent*—leaves room for the wedge of F. Lee

Bailey's defense strategy, which took advantage of the only uncertainty that could persist in the face of so much evidence, the presence of intent or willingness. To call into question the incontrovertible evidence of Hearst's words and deeds, Singer and Lifton offered forensic analyses of audio recordings and video footage of Hearst during her captivity. Through interviews and these texts, both psychologists were tasked with analyzing the defendant's mental state in her SLA captivity. In her scripted communiqués, the defense's expert witnesses concluded that Hearst had not written the documents in question and, later, that the language she used was not "her own." The examinations discuss the "canned" quality of this language as it conforms to the linguistic criteria of coercive persuasion, and they go on to dismiss the contents of that language as symptoms of an illness rather than as viable political statements.

The task of the brainwashing defense, then, was to make Patty's participation in the SLA the result of "traumatic neurosis." Singer describes how she ascertains (correctly) that Hearst's words in the communiqués are not her own, first concluding that the speech has been scripted, from noises of pages turning to the "breathing patterns," the "groupings of words," and the automatic manner of Hearst's speaking, which is indeed an eerie monotone.[36] In one of Hearst's statements read aloud in court, the SLA explains, for instance, how "expropriations" are "revolutionary" rather than "criminal act[s]," a definitional game that shifts the context of their actions to those of unacknowledged legislators.[37] Singer claims that the words are not in Hearst's own "style," marking them as part of a reeducation campaign. The defense makes much of the intangible "style" of language and particularly of its "canned" quality:

SINGER: I used both those terms, Mr. Bancroft, canned language and loading of the language.

BANCROFT: All right. Now, did you find those expressions in the canned and loaded language in the SLA tapes particularly inconsistent or uncharacteristic of the defendant Hearst in her expressions that you knew about prior to February 4, 1974?

SINGER: In a number of those SLA tapes they are highly inconsistent. . . . There are Maoist phrases and politicized interpretations of things

FIFTY CENTS

APRIL 29, 1974

TIME

Patty Hearst

Patty Hearst on the cover of *TIME* magazine, April 29, 1974. Inset within the portrait are a photograph of the heiress released by the Symbionese Liberation Army, *left*, and a frame from the closed-circuit camera footage of Hearst captured during the Hibernia Bank robbery for which she would be tried in 1976.

that didn't at all seem characteristic of her style, and there were catch phrases, like Muerta Venceremos [i.e., *Patria o muerte! Venceremos!*]; and it was my impression from having interviewed her and talked to her about such matters that these represented a rather classical use of canned phrases as they are instilled coercively into victims of a thought reform or coercive persuasion process.[38]

In Singer's analysis, these canned phrases are parroted ideas, not truly processed but instead instilled in Hearst by means of coercive persuasion. Even the words and phrases that had not been literally scripted for Hearst in the recordings were *canned*. Singer does little to clarify precisely how such coercive instillment of canned language has worked with Hearst. The idea suggests comparison, though, to *The Manchurian Candidate*, in which Captain Marco repeats the phrase "Raymond Shaw is the kindest, bravest, warmest, most wonderful human being I've ever known in my life" as the extreme of the canned language to which Singer refers. Singer's views on language follow closely alongside George Orwell's, which, as discussed in chapter 1, attributed to set phrases and terms like "Comintern" the possibility of parroting and not thinking through the full meaning and history of a word. The jargon of revolution certainly presents the most distinctive characteristic of Hearst's language, and it was at the very least conclusive that her style of speech had been changed by her captivity with the SLA. A similar discussion occurred in Lifton's later testimony and cross-examination, which also cites Hearst's captivity in the closet and the SLA's "canned language" as criteria for coercive persuasion and traumatic neurosis.[39]

When Ted Patrick visited the Hearst family to offer his deprogramming services in 1974, his analysis was strikingly similar to Singer's and Lifton's. Patrick compared the SLA's techniques with those of the Unification Church, and then he brought his cassette recorder to play a tape of a young woman in the Hare Krishna. The young woman on the tape denounced her parents in a voice that was "frail, empty, hollow, and without expression," and the Hearsts agreed that the young woman's voice had a quality quite similar to their daughter's.[40] Much like the cinematic situation of viewing an automaton character's empty and robotic

motion, the tape recordings of Hearst, which are indeed flat and haunt-
ing, harbor the threat of being devoid of subjective intention. Ted Patrick
described her mental state as that of "a psychological vegetable."[41]

In a curious twist, while they had been on the run in the year and a
half prior to Hearst's arrest, the SLA themselves had leveraged their own
"brainwashing defense," taking up the media's frequent use of "brainwashing" in descriptions of the SLA's tactics. In the document gener-
ated by Hearst's comrades while they were on the run, they write, and
Patty says, "I feel that the term 'brainwashing' only has meaning when
one is referring to the process which begins in the school system and
is continued via the controlled media, the process whereby people are
conditioned to passively take their place in society as slaves of the ruling
class. . . . I've been brainwashed for twenty years, but it only took the
SLA six weeks to straighten me out."[42] When Singer is cross-examined
about this particular recording, an irony goes unremarked: in calling
Hearst's denial that she had been brainwashed, too, part of the coercive
persuasion's uncharacteristic "style," Singer lets the content of the tape
pass without comment.[43] The SLA repeatedly insisted on Hearst's freedom, particularly in the widely cited communiqué that ends with the
announcement of her free choice to join the SLA: "I have been given
the choice of (1) being released in a safe area or (2) joining the forces
of the Symbionese Liberation Army and fighting for my freedom and
the freedom of all oppressed people. I have chosen to stay and fight."[44]
Brainwashing, which the SLA had called "ridiculous" "beyond belief" as
an accusation in one communiqué, becomes one of their most powerful
rhetorical weapons when it suits their purposes.

Does that reversibility, then, wholly evacuate the brainwashing discourse of any positive content? As the question played out in the trial,
the "expert" role of the psychological testimony was never wholly beyond
question. Expert witnesses for the prosecution contradicted Lifton and
Singer, and in his closing argument, the prosecutor, James Browning,
described that whole portion of the long trial as a "wash transaction," in
which one side canceled out the other.[45] And the brainwashing argument failed to persuade the jury of Hearst's innocence. Among the several reasons for this outcome is the representational burden placed on

scientific explanation. All the expert witnesses had to deal with the impossibility of *seeing* Hearst's volition, or lack thereof, in her actions. There was no visible sign of her lack of *willing* participation in the crimes, despite the abundance of evidence provided by her own recorded words and gestures, the audiotapes and videotape on which Hearst's voice and moving image were captured. Instead, the frequently linguistic criteria for "traumatic neurosis" offered an ersatz medical standard for deciding whether Patty was acting freely.

In the video recording that showed Patty's participation in the bank robbery, however, the seed of uncertainty had been firmly planted. Browning's closing remarks admit as much, and he implores the jury to find her *willingness* captured within these images: "There is no question but that she is depicted in those motion pictures [of the Hibernia Bank robbery]. And you have seen them yourselves, you have seen how she looks in these motion pictures. You are entitled to consider how she looks in making up your mind as to whether she was in that bank apparently voluntarily."[46] As captured on camera, Hearst's will is as invisible as those of the body snatchers or Raymond Shaw, and it remains remarkable that a case for Hearst as a human automaton, in recorded speech and action, could have been taken at all seriously.[47]

Hearst's own memoir, *Every Secret Thing*, which she cowrote with Alvin Moscow in 1981, rehearses this automatism, this time as a feature of written narrative. Hearst's memoir is notable for its insistent renarration of a partial conversion, where her volition and willingness in her own actions recede from view as much as they had in her trial. She describes, for instance, the idea of having "acted instinctively" from her training as a "soldier" that "caused" her to fire a gun in the botched robbery of Mel's Sporting Goods store, an action that had counted against her in the trial for the bank robbery.[48] Likewise, when she describes herself from others' perspectives, she often suggests, but never conclusively, that her state is not an ordinary one: she notes that Wendy Yoshimura described her later as "a dull vegetable without a mind of [her] own."[49] In sum, she repeatedly makes use of the criteria for coercive persuasion in the trial, including the linguistic criteria discussed by Singer: the SLA members "spoke in slogans," Hearst writes, and the "revolution was their religion and they were its fanatics."[50] The memoir becomes a compelling narrative, then, by stringing the reader along with bits of evidence in this same vein, never quite describing whether Hearst believes or simply goes along with the SLA's practices. Toward the end of the text, Hearst offers the account of her conversion as a not-quite-narratable coercive persuasion:

> At first, I remembered, I thought I was humoring [my kidnapper] Cin[que] by telling him what he wanted to hear. . . . I thought at first that they all were stark raving mad, crazy, out of their minds, but in time I came to accept what they believed because of the repetition of those daily criticism/self-criticism meetings. It is a difficult process to understand. We are raised to believe we enjoy freedom of thought. Cinque and the others told me that the people of America were all brainwashed in believing in bourgeois principles, ranging from family and monogamy to capitalism and two big cars in the garage. . . . It was all a question of who was brainwashing whom. But the single difference in my own case: I had been persuaded coercively—by force; no one had forced the Harrises or the [other members of the SLA] to adopt the SLA principles and Codes of War.[51]

In the final instance, here, Hearst takes recourse to the criteria laid out by Singer and Lifton in the trial. She had indeed been persuaded by the SLA, but her narrative goes to the same pains as the scientists had done to claim that she had been "persuaded coercively." Hearst has often mentioned that the lasting ideological change she made during the SLA experience had been adopting a feminist political stance. But in an era in which feminist thought and life writing so often focused on the recovery of the "voice," of reclaiming agency and self-determination, Hearst's memoir, in using this scientific discourse, succeeds in effacing her own agency from the text. While the antitotalitarian writings that gave us the term *brainwashing* had provided the SLA with some of its language of defiance, the scientific criteria that had been used to define brainwashing provided Hearst with the means of depicting herself as a human automaton.

Cult Panics and Cool Cults

The story of postwar cults, then, has raised questions about the visibility of individual agency, about the determining dynamics of social space, and about the roles of scientific expertise in explaining apparent lapses of free will. Such incidents prove fascinating and even entertaining on multiple levels, and stories about cults continued to proliferate long after the most shocking events of the 1970s and 1990s occurred. Hearst's story, like cult narratives generally, has circulated widely and in a variety of forms, with a new novel, poem, documentary, or popular history appearing steadily every two to three years since the event. It is to the circulation of these cult representations that I turn for the remainder of this chapter. The cult has been used more and less seriously, and with different goals in mind, for audiences all over the social map: didactic fiction to keep children and young adult audiences from joining cults; postmodern literary fiction that positions the cult as emblematic of contemporary mass culture; historical fiction that sees cults and Patty Hearst as emblematic of the 1970s in particular; exposés of particular religious groups *as* cults; and reportage, nonfiction, and drama that depict the cult as a worrisome, if titillating, space in which individuals' autonomy may be seriously threatened. Like the phenomenon of automatic movement in film, the idea of the cult has circulated with particular ease between so-called high and low cultural forms. As a problem for human programming, the cult has been considered with both the urgency of a mass media moral panic and with the cool remove of literary fiction by Don DeLillo, Chuck Palahniuk, Katherine Dunn, Christopher Sorrentino, and many others. Between them, and despite the differences in tone, the questions of structure and agency remain remarkably constant, as do the conventions of the *topos* of the cult.

Hearst's trial has been the subject of many narratives that envision Hearst as an icon of the American 1970s or of the contemporary era in U.S. culture more broadly. Conversion—like the recognition and reversal that constitute it—is nothing if not the stuff of literary narrative, and the question of Patty Hearst's free will and the broader social meaning of her actions has been recast in different narratives almost nonstop since the end of the trial, in memoirs of Hearst, her fiancé, her collaborator, and in poems, films, and, with renewed vigor since 2001, novels. Susan Choi, in the novel *American Woman*,[52] a 2003 Pulitzer Prize finalist, looks back at the episode through the eyes of Hearst's collaborator-on-the-run Wendy Yoshimura, a cause célèbre in the Asian American community, with an eye on the transition between the political cultures of the 1960s and the 1970s. Pamela White Hadas likewise includes Hearst in her book of poetry, *Beside Herself: Pocahontas to Patty Hearst*,[53] as the end point in a prosopography of American women. Hadas's poem "Patty Hearst: Versions of Her Story" emphasizes in fairy-tale style the various conflicting versions of her story, but with Hearst's own voice formally suppressed, appearing only as quotations within a reporter's version of the story.[54] From the mid-1970s onward, figures associated with Hearst published memoirs, including her friend and U.S. marshal Janey Jimenez and her former fiancé Stephen Weed. Paul Schrader's feature film *Patty Hearst* (1988) and the PBS documentary *Guerilla: The Taking of Patty Hearst* (2004) both make much of the Hibernia Bank robbery video footage and the eerie recordings of Hearst's voice. Among these should be counted the historian William Graebner's *Patty's Got a Gun: Patricia Hearst in 1970s America,* which, as a popular academic history, uses Hearst's trial as a window into the culture of the era.[55]

Christopher Sorrentino, in the National Book Award finalist novel *Trance*,[56] gives perhaps the most sophisticated and sustained fictional treatment of Hearst's narrative. Sorrentino makes the most of the literary and stylistic forms that make up the *topos* of the cult, as well as the emphasis on slippery interpretation between worlds that must take place when we translate between the specialized language of the cult or the revolutionary cell and ordinary America. The narrative, like *TIME*'s coverage of Ted Patrick, connects the SLA to the widest possible variety of scientific paradigms and cultural forms, from brainwashing to *Rosemary's Baby* (Roman Polanski, 1968). Gabi, an SLA member, declares that the SLA "is turning into like one of those whatchamacallits I read about in *TIME* last year. Cults."[57] Her comrade Zoya responds as though she were talking about *Rosemary's Baby*: "Like people in hoods and altars? Drinking blood? You insult . . . our hard work, our comrades." Gabi responds, "'Like the Hare Krishnas. The Moonies. Utopian hucksters," an exchange that accurately registers the novelty of cult discourse at the time as well as its intuitive applicability to the SLA situation. Sorrentino also listens in on a conspiracy theorist who imagines that SLA member Donald DeFreeze has been "programmed . . . via electrodes implanted directly in his brain" to work as "an element of a new CIA program" to "infiltrate leftist groups."[58] In this range of forms, Sorrentino emphasizes the variety of interpretive possibilities that surrounded Patty Hearst and that seem often to swirl, confused, around her in the narrative. Sorrentino's narrative most successfully conveys that confusion through formal experiments, such as the following that combines dialogue, news broadcasts, radio, and police megaphones in play-by-play reactions to the first SLA shootout that resulted in the deaths of many of the members:

> *Police are saying the fugitives are better armed than they are.*
> *holding the Negro residents of the house hostage.*
> Teko: "Bullshit! Fascist bullshit!"
> *here in the newsroom we have a noted expert*
> *We have been informed that more than three hundred police and FBI agents are participating in this operation, an awesome amount of firepower marshaled against the radical sect the SLA.*
> *who is here to tell us about the SLA and their strange beliefs.*
> Reporters and police both fall back, fall back in a wave broadcast in a series of shaky images by the MiniCam Unit of KNXT-TV
> The dry wind sends the CS [tear] gas coming back. Searing and choking.
> FALL BACK, GET BACK.[59]

Even during the midst of police action, expert and media perspectives paint the SLA as a "sect" with "strange beliefs." Sorrentino's novel plays at a polyvocal aesthetic to juxtapose different registers of discourse attempting to home in on the subject. While the scene points out the SLA's dissenting interpretation of the situation at hand, it also points out that their social space is literally a building on fire, viewed as a den of crime and lunacy by all those who circle around outside. While Sorrentino's novel largely embraces the relativism that might consider the SLA's point of view as legitimate—to view them, as the sociologists of NRMs do, as a culture—it also occasionally punctures that relativism with humor at the SLA's expense. The character Guy, for instance (a pseudonym for Jack Scott, who helped transport Hearst and others when the SLA was on the run), offers a nonrevolutionary progressive perspective: with the SLA members, "an entirely new subset of clichés was coming of age. He hadn't realized the SLA took its rhetoric literally."[60] Hearing the repeated and ultimately nonsensical accusations of "fascism" in the SLA's conversations, "Guy thinks, fascist fascist fascist fascist fascist," and later, he compares their "Symbonia" to the Marx brothers' "Freedonia."[61] It is on the basis of linguistic criteria—what Singer had called "canned language" in Hearst's trial—that Guy ultimately refuses to take the SLA seriously. However naive or misguided, the SLA offers a way of seeing and interpreting the world radically differently, an American 1970s in which revolution could be imminently possible. The *topos* of the cult in this sense both protects the members from the ridicule of the outside world and makes them the subject of that ridicule.

As the 1970s wore on and these cults grew more and more widespread, mainstream media consideration of them took on a panicked tone. The news coverage of the November 1978 mass suicide at the People's Temple, a California-based religious group that had relocated to the South American republic of Guyana, reached the height of the moral panic that surrounded cults.[62] A series of articles in *TIME* showed its readers the exotic locale of the People's Temple, the strange appeal of the cult leader Jim Jones, and the broader context of cults as a social problem. One such story, "Follow the Leader: How Cults Lure the Drifting and Discontented—and Keep Them," describes how "cults such as Sun Myung Moon's Unification Church, Scientology, Synanon, Hare Krishna and Children of God offer a refuge from the storms of the world."[63] The article describes the process of "'programming' or 'brainwashing'": the "message is incessantly drummed in," "ties are severed" with friends and family outside the group, and devotion to a "father figure" is encouraged. Such descriptions were in keeping with sociological and psychological criteria of the *topos* of the cult. The article interviews an ex–Hare

Krishna member, who describes the chanting program, beginning daily at 4:00 A.M., as an experience in which "you're not really there."

The flurry of discourse surrounding the devastating events at Jonestown culminated in a Senate hearing on cults in February 1979, led by the senator (later presidential candidate) Robert Dole. As *TIME* reported, John Clark, professor of psychiatry at Harvard, said these cults "raised frightening specters of suicide, 'uncontrolled violence,' and . . . mindless zombies who pose a clear threat to democratic societies."[64] Both the science fictional imagery and the rhetoric about threats to democracy pervade anticult discourses from the tabloids to Senate hearings. Members of the Unification Church protested outside, as primarily anticult witnesses recommended courses of action to curb cult recruitment. The Unification Church had recently won litigation against cult deprogrammers. Efforts to curtail NRMs would run against the religious freedoms afforded by the First Amendment, in a country whose entire history had been marked by religious revivals and Utopian sects. Clark's language about the "mindless zombies" in cults was matched by the American Civil Liberties Union's (ACLU's) representative, Jeremiah Gutman, who testified before the Senate that "'forced psychotherapy' to attack unwanted belief is 'precisely what is going on in the Soviet Union today and precisely what Ted Patrick does on a smaller scale. It is already against the law.'"[65] The same interpretive impasse that has since beset sociologists of NRMs—is it a social problem to be fixed or a form of diversity to be respected?—emerges within that rebuttal from the ACLU, with a dash of anticommunist rhetoric for good measure.

Literary postmodernists like Sorrentino often engage with the religious or political cult, but that engagement is most frequently marked by a cool remove from the moral panic or the sentimental conventions that mark didactic fiction, news reportage, and documentaries. Amy Hungerford has argued that postmodern fiction often traffics in a "mystical understanding of language" that has been shuttled in from more traditional forms of religiosity.[66] Where Hungerford makes a convincing case that a certain fullness of language itself replaces religion in, for instance, Toni Morrison's fiction, I would contend that postmodern fiction on cults can also play on language's potential, in the case of cult jargon, to be simultaneously overfull and utterly empty. Postmodern fictions of the cult tend to emphasize images both of the cult believer as a dupe and of belief itself as a linguistic construction. One often finds, too, in the depiction of the cult, echoes of the dupe to the consumer culture against which postmodern literature often poses itself, even as novelists revel in the raw linguistic material drawn from cults. Chuck Palahniuk's fiction, for instance, employs the figure of the cult as a curiosity, one set up by the charlatan narrator of *Survivor* and another in the revolutionary political enclave in *Fight Club*.[67] In the latter, a society begun by rejecting the emptiness of contemporary consumer culture, Tyler Durden's devotees find themselves acting as equally mindless followers of a revolutionary agenda. [68]

Don DeLillo's novel *Mao II* concerns itself with the conceptual problems of the cult and language as well, in an exemplary examination of scenes of mass consciousness and of enclosure in post-1970s U.S. culture. The cult serves as a means to explore the fate of the individual within mass culture in DeLillo's novel, in both individual and historical terms.[69] The novel's opening scene depicts one of the famous mass weddings of the Unification Church, led by Reverend Sun Myung Moon. This scene is narrated mainly from the perspective of the principal character Karen's parents, who were invited to the wedding but watch the ceremony through binoculars from the stands at Yankee Stadium. Rather than the sequence of forms of consent that begins with the father giving away the bride, the mass spectacle gives the wedding vows an eerily automatic tinge. In a passage soon afterward, Karen describes her closeness to Reverend Moon: "They know him at a molecular level. He lives in them like chains of matter that determine who they are. This is a man of chunky build who saw Jesus on a mountainside."[70] Karen knows her parents doubt the authenticity of her belief in Reverend Moon, and she announces, in her thoughts, the rhetorical effectiveness of nomenclature: the "word is 'cult.' How they love to use it against us. Gives them the false term they need to define us as eerie-eyed children."[71] Karen later narrates a scene of forceful, Patrick-esque "deprogramming," which rehearses the contradictions inherent in the practice of reverse brainwashing.[72]

This young woman goes on to live in an isolated house with the novelist Bill, a figure whose charismatic leadership and devoted assistant Scott invite an implicit comparison to the cult. That comparison elicits the problem that Lifton and other sociologists took up, that of the limits of the cult form within social organizations. Indeed, the broad casual usage of the term suggests that any form of unusual devotion to a leader, whether he be a favorite reclusive author or Chairman Mao, bears comparison, by degree, to this social formation. The situation that encloses the three characters in the author's cabin creates a comical *reductio ad absurdum* of this logic. Scott, for instance, often ends the expression of complex ideas with "Quoting Bill," as though he were never thinking for himself but always deferring to the teachings of the master.[73] The last we see of Karen in the novel, after she has lived in seclusion with Bill and Scott, is that she goes back out onto the street, proselytizing for the Unification Church and echoing Robert Jay Lifton's emphasis on totality in social forms: "We are protected by the total power of our true father. We are the total children. All doubt will vanish in the arms of total control."[74] Karen finds the cult beautiful, which, in the postmodern and ironic mode of DeLillo's characterization, signals an idea to be considered at cool remove, even when the last the reader hears of the character is that "she had Master's total voice ready in her head."[75] (Scott's final words in the novel, "Quoting Bill," also reflect the ever-presence of Bill's voice in his follower's head.)[76] Near the novel's end, a photographer interviews a terrorist, who tells her that "Mao believed in the process of thought reform" and, in the next breath, that "there is a desire for Mao that will sweep the world."[77] The novel's obsession with the consciousness of the future—a future that "belongs to crowds"—imagines the figure of cult consciousness as a sort of conceptual threat within postmodernity.[78] The emptying-out of the individual in mass culture finds the cult as the emergent social form of DeLillo's postmodernity, while the individual, and in particular the novelist, stands as a sort of relic, an artifact from industrial modernity. Sun Myung Moon, although his role in the novel is a minor one, is the leading candidate for the *Mao II* of the novel's title, which refers also to Andy Warhol's portrait of the Chairman. That logic of replaceability between the cult and the totalitarian state, and of the essential emptiness of the leaders in the face of the emptying-out of belief itself, structures the novel from the title forward. That association between the cult and postmodernism meshes well with Nancy Isenberg's assessment of Patty Hearst as "perhaps the best of all postmodern subjects."[79]

But in historical terms, the political applications of this idea of the cult have also fluctuated since the 1990s, the end of the Cold War, the advent of what Michael Hardt and Antonio Negri have referred to as "Empire," and the beginning of the War on Terror. In 1993, when the Branch Davidian compound burned down and, by coincidence, the first World Trade Center bombing was orchestrated by the Muslim extremist Khaled Sheikh Mohammed, *TIME* magazine ran a cover story comparing the two events.[80] The author, Lance Morrow, turns to Francis Fukuyama's work on the "end of history," wherein the fall of the Soviet Union represents a triumph of liberal capitalist democracy, and where, according to Morrow, the "world's civilization would settle upon a kind of sunsplashed plateau of democratic pluralism and free-market rationalism."[81] The Waco and World Trade Center events revealed for Morrow a "dark, chaotic side" to the end of history, wherein the "collapse of the binary cold war configuration has produced an unstable free-form arrangement of forces and impulses loose in the world, often traveling forward or backward at high historical speed."[82] To Morrow, the "micro-fanaticism" of the Branch Davidians and the "macro-drama" of Islamic extremism—the "repugnance and envy that often resolves itself into militant fundamentalist anger"—are two sides of the same coin.[83] Here, fundamentalism is a symptom of a reaction against global modernity, one that grows within cult-shaped enclaves of refusal and delusion. This argument would anticipate those of Samuel Huntington's 1996 book *The Clash of Civilizations and the Remaking of the World Order* and Benjamin R. Barber's 1995 book *Jihad vs. McWorld*, both of which would reenter wide circulation in the United States immediately following September 11, 2001.[84]

CHAPTER FIVE

Fundamentalist Automatons

Representing Terrorist Consciousness
in the War on Terror

"**B**y the time you watch this, you'll have read a lot of things about me. About what I've done. So I want to explain myself, so that you'll know the truth. . . . I was beaten. And I was tortured. And I was subjected to long periods of total isolation. People will say that I was broken. That I was brainwashed. People will say that I was turned into a terrorist, taught to hate my country. I love my country. What I am is a Marine."[1] So claims Nicholas Brody, the protagonist of Showtime's *Homeland*, in a video recording made before an abortive suicide-bombing attempt. Within the show, Brody here assumes that the media will refuse to understand his act and denounce the bombing as the product of an unsound mind—a hater of America, a "broken" man, or a "brainwashed" man—and an enemy whose motivations can be safely ignored. He resists this pathologization of his act by recording this video, which will in turn be the document that ultimately incriminates him. His assumption that he would be called "brainwashed" hearkens back to figures like John Walker Lindh, José Padilla, Richard Reid, Zacarias Moussaoui, and others who, as I discussed in the introduction to *Human Programming*, were also called "brainwashed" in the news media by baffled friends and family. As if to supplement Brody's statement here, the actor who plays him, Damian Lewis, also felt compelled to deny Brody's having been brainwashed; in a promotional interview with Jeremy Egner, Lewis insisted that "Brody chooses to act; he's not brainwashed into doing something," even as the tension between choice and brainwashing carries much of the suspense in *Homeland*'s first season. Among the various paradigms in which terrorists have been represented in contemporary culture—as neomedieval savages or as the nebulously racialized group of "Muslims, Arabs, and Middle-Easterners"—the scientific paradigms and cinematic images of human automatism are regularly employed when texts delve into the question of the terrorist's psychological state. Across many facets of intellectual and cultural production, American representations of Islam are suffused with scientific ideas about, and the aesthetic conventions of, fanaticism and discipline that developed during the Cold War. These include the *topos* of the cult, the metaphor of brainwashing, and even the metaphors associated with computer programming.[2] In the War on Terror, the fundamentalist terrorist takes shape as the most recent in a lineage of programmed avatars of un-American "unfreedom."

At the same time the human automaton has been used to disavow or pathologize terrorists and fundamentalists, more sympathetic representations of the terrorist, such as *Homeland,* have developed a complex relationship with that "brainwashing" and automaton imagery in popular culture and scientific discourse. A set of self-consciously high-cultural texts, including twenty-first-century serial television, novels, and films, has used the figure of the terrorist to probe the limits of Lionel Trilling's "liberal imagination," that is, the ability of authors to imagine, and for readers to grow more capacious by reading, characters and kinds of consciousness that are radically different from one's own. Contemporary readers have come to trust fiction as a medium for coming to understand others' experiences and states of consciousness, an assumption I want here to take as a phenomenon in its own right as I explore how these texts work with and against the conventions of human automatism. As other instances of the human automaton have done, these fictional representations of terrorist consciousness take great interest in the tension between scientific knowledge and ethical acknowledgment, and both the writers and critics of these texts can be seen searching for the criteria that constitute a successful sympathetic understanding of such a radically different consciousness. The stakes of this success, moreover, are high in U.S. literary culture: *Homeland* won four Emmy Awards for its first season at the same time it was called "TV's most Islamophobic show,"[3] for instance, while Don DeLillo's *Falling Man* has had at least fifty-nine academic articles published on it in the years since its publication.[4] As I claim in this chapter, these texts cultivate a complex understanding of the scientific discourses and fictional conventions of human automatism to which Americans often resort in representing fanaticism, torture, and other extreme mental and social phenomena. This takes place, as I'll show through discussions of Don DeLillo's *Falling Man*, criticism of John Updike's *Terrorist*,[5] and the first season of *Homeland*, as a pattern of sympathetic pathology. This pattern, the strikingly common use of American protagonists with mental disorders in fictions about terrorism, emerges as a mechanism through which these texts compare the presumed mental freedom of ordinary Americans to the supposed mental unfreedom of fundamentalist terrorists. That pattern of using pathology in fiction is a novel one with respect to the conventions of 1960s antipsychiatry literature, and, paradoxically, these texts encourage sympathetic understanding of others through scientific and mechanistic understandings of ordinary characters' minds. As such, these explorations of sympathetic pathology constitute some of the most sophisticated and forward-looking uses of the human automaton in U.S. culture.

Abstractive Pathology: Muslim and Machine

By considering how fundamentalist terrorists are represented according to the paradigms of automatism, I hope to complement other useful accounts of racial and religious representation that have emerged since 9/11. The forms of discrimination and racism that emerged to meet the supposed terrorist threats to the United States since 2001 have been drawn from a wide array of discourses, each adapted to distinct rhetorical and practical uses. Bruce Holsinger, for instance, has convincingly argued that the War on Terror has been suffused with discourses of "neomedievalism," wherein Islamic fundamentalism can be cast as a constitutively antimodern and peculiarly indefensible and ignorant way of life in the contemporary world.[6] This is borne out particularly well by George W. Bush's "crusade" rhetoric and the recirculation in diplomatic circles after

9/11 of Raphael Patai's 1973 *The Arab Mind,* a text whose focus is often on an atavistic "pre-Islamic heritage" with features like *"lex talionis,"* the "imperative of preserving one's 'face,'" and an emphasis on killing for group or family honor.[7] With reference to new forms of surveillance, profiling, and police discrimination that have emerged, Leti Volpp has described a "racialization" of Islam, where the ersatz grouping of "Middle Eastern, Arab, or Muslim" citizens emerged as a slippery racial category in the immediate aftermath of 9/11.[8] Jasbir Puar, for her part, has also convincingly argued that Arab and Muslim bodies are often imagined as constitutively queer, in a wide-ranging extension of the gender dynamics Edward Said associated with Orientalism.[9] Complementing Puar's, Volpp's, and Holsinger's influential observations about the categorical complexity of Muslim and Arab American identity after 9/11, Leerom Medovoi's description of "dogma-line racism" allows us to consider the extent to which Islamophobia also constitutes a fear of, or fascination with, mental unfreedom. For Medovoi, a key feature of the "dogma line" (as opposed to the "color line") is that, rather than being judged in terms of racial inferiority, the Muslim is judged as an enemy primarily through her fanaticism.[10] This concept makes visible a form of discrimination, "in primary reference to mind rather than body, ideology rather than corporeality," that is, "according to theologies, creeds, beliefs, faiths, and ideas, rather than their color, face, hair, blood, and origin."[11] It is through this emphasis on mind and belief that new forms of discrimination manage, even across racial lines, to draw on scientific discourses of human programming that have developed since World War II. While Medovoi traces this "dogma line" in a centuries-long arc alongside the roles of religion in the public sphere, the popular understanding of fanaticism in the post–World War II period has also borrowed significantly from scientific discourses. The sociology and psychology of cults and NRMs, and public-sphere metaphors of programmability and brainwashing alike, have provided accounts of the subhuman or machinelike mind that can espouse a radical or unthinkable idea. While what Volpp calls the newly "racialized" group of "Middle Eastern, Arab, or Muslim" has proven indispensible to our understandings of, for instance, racial profiling since 9/11, the dynamics of "dogma-line racism" describe forms of public-sphere rhetoric that are even more flexible and broadly applicable, such as the media depictions of supposedly "brainwashed" fundamentalists like John Walker Lindh, Richard Reid, and others outside that pseudo-racial group.[12] And, even more particularly for my purposes, in public-sphere modes where the individual psychology of the terrorist is at question, that dogma-line racism is often suffused with the scientific and pseudoscientific discourses I've been tracing throughout *Human Programming*: the sociological *topos* of the cult, the psychology of coercive persuasion, and various informal accounts of behavioral conditioning as "brainwashing." Moreover, this pattern regularly follows into investigations of terrorist consciousness in television, film, and literary fiction.

Post-9/11 texts that represent Islam through this dogma-line racism combine the scientific features of Cold War psychology and sociology with much longer traditions of anti-Muslim prejudice and sentiment. Europeans and Americans have associated Islam with fanaticism for many centuries; a link that has, since 9/11, often been tacitly strengthened by Cold War science. Alberto Toscano's recent philosophical genealogy of fanaticism shows that this link between Islam and fanaticism, an Orientalist foil to Enlightenment reason, runs deep in the Western philosophical tradition.[13] As fanaticism became an important conceptual problem in the Enlightenment, Islam was often called upon—as it was in Voltaire's 1741 play *Fanaticism, or Mahomet the Prophet*—as an emblem of fanatical devotion.[14] Toscano identifies a homology in Hegel's world-historical philosophy between the account of Muslim belief and the famous analytic of the "terror" in the French Revolution, as an "enthusiasm for the abstract."[15] As a supposedly "violently intolerant" religion, Hegel's Islam resembles Hegel's French revolutionary terror as a form of intellectual immaturity, a faith in a form of universality that can "bear only a desolating destructive relation to the concrete."[16] This association between fanaticism and Islam persists into the mid-twentieth century, particularly in Jules Monnerot's *Sociologie du communisme* (1949), which claims that "communism was the Islam of the twentieth century," an assertion that Hannah Arendt takes to task in her 1953 essay "Religion and Politics."[17] Throughout this intellectual–historical career of "fanaticism," Islam has remained an exotic religion onto which anxieties about the powers of both Oriental despotism and religious dogmatism have been projected, and, since the French Revolution, it has been explicitly contrasted with pragmatic and just forms of society. With both Hegel's characterization and Monnerot's declaration that "communism is the Islam of the twentieth century," the religious content of Islam is all but evacuated, and the political signifier of "Islam" stands in for the function of false, fanatical consciousness.

In the present, scientific and governmental expertise on Islam blends these ideas about fanaticism with other tendencies of contemporary ethnography. Patai's *The Arab Mind,* for instance, an Orientalist text that regained currency after 2001 in U.S. governmental circles, mixes an account of a supposed Arab racial atavism with accounts of the "religiocentrism" of Muslims.[18] In a related arm of public discourse, popular nonfiction books, such as Samuel P. Huntington's *The Clash of Civilizations and the Remaking of World Order* and Benjamin Barber's *Jihad vs. McWorld,* quickly recirculated after 9/11, frequently cited for their titular simplifications of the differences between the liberal Western self and the neomedieval fundamentalisms that refuse contemporary secular capitalist values. This set of popular and expert discourses on terrorism set out to use scientific authority and expertise drawn primarily from sociology and political science to understand the terrorist as a stable and manageable category.

In opposition to those appeals to ethnographic expertise, the term *Islamophobia* has been employed in the present decade as a way of describing discrimination against Muslims and as an alternative name for the dogma-line racism that Medovoi identifies. More than a dozen books have appeared with "Islamophobia" in their titles, along with at least one far right retort using this newly popular term against the grain to argue that Islamophobia is entirely justified and that the liberal pieties of tolerance are also totalitarian.[19] Islamic studies scholars John L. Esposito and Ibrahim Kalin, for instance, in their *Islamophobia: The Challenge of Pluralism in the 21st Century,* document in case studies instances and dimensions of discrimination against Muslims in the United States, one of which is the "perception that the religion of Islam has no common values with the West, is inferior to the West, and that it really is a violent political ideology rather than a source of faith and spirituality, unlike the other Abrahamic religions."[20] In identifying this perception, Esposito and Kalin describe a strategy of imagining Islam as a source of violence—a social problem, to be viewed and solved from a distance—rather than as the source of individual spiritual experience that a more sympathetic perspective might recognize.

This set of perspectives on Islam has, lastly, suffused public-sphere discourse on terrorism over the past decade, a sphere in which the problem of pathologization in expert knowledge becomes clearest. As Lisa Stampnitzky has argued, it was during the mid-1970s that "terrorism" was consolidated as an academic problem and subject of social–scientific expertise, and, like Patai's *Arab Mind,* that work has circulated with renewed vigor since 2001. Stampnitzky's study of the history of that expertise on terror illuminates its central paradox: "since 9/11 Americans have been told that terrorists are pathological evildoers, beyond our comprehension," but that they are at the same time the objects of systematic study by many governmental and nongovernmental agencies.[21] Writing about the news coverage of suicide bombers in recent decades, Jacqueline Rose has claimed that "the argument that suicide bombers should not, or cannot, be understood" rests on a "subtext of dehumanization."[22] In dehumanizing or pathologizing the terrorist suicide bomber, one not only succeeds in invalidating the meaning and the motivations of the act but, more important, places the terrorist into a too-comfortably distant and self-contained category. Jasbir Puar notes, too, that the complex stories about terrorism and its multifaceted social organizations and social causes are often reduced into "the story of individual responsibility and individuated pathology," that is, simple explanations based on moral wrongs and diseased minds.[23] In focusing on individuals and the representation of individual psychology, however, fictional texts often work within this horizon of meaning making. Such a representation seems particularly difficult for fictional texts, and as I explain in the following pages through the examples of *Battlestar Galactica, Falling Man,* and *Homeland,* a consistent new strategy is one that dehumanizes ordinary American characters in the course of attempting to represent and understand the terrorist.

Along these lines, documentary and nonfiction representations of terrorists often resort to pathology, explaining the actions of Muslim terrorists as a form of madness, compulsion, or programming. Take the 2010 documentary *American Jihadist,* a profile of the Washington, D.C., born African American freedom fighter Isa Abdullah Ali.[24] Ali has worked in Beirut and elsewhere in the Middle East for Islamist causes, primarily as a soldier and assassin. In keeping with the narrative conventions of the criminal profile, the documentary ranges in its explanations of Isa's behavior, from a neo-Freudian focus on the impact of childhood trauma to the enabling role of Islamist rhetoric, as the kind of "violent political ideology" that Esposito and Kalin identify in their work.[25] A CIA profiler interviewed in the film thinks of Islam itself as an arbitrary outlet for Ali's psychological violence, and he posits that "the cause is an exoskeleton"; the same profiler imagines Ali's language as canned, calling it "boilerplate" that "sounds good in front of a camera." That explanation of canned language, as in *1984* and the expertise on cults discussed in previous chapters, allows the profiler to discount the content of Isa's speech, and it encourages the viewer to do the same. Although the documentary gives Ali his own voice and is structured as a variety of explanatory perspectives on the man, the expert perspectives on him transform his own accounts of himself into symptomatic speech, empty rationales for an uncontrollable, pathological tendency to violence. This contemporary Islamophobia is defined centrally through the symptomatic reading of Islam itself as a form of socially or psychologically rooted false consciousness. It is a false consciousness that is at once mechanical, inauthentic, and unsettlingly unfree.

Such a pathologizing tendency also carries into some literary fiction that addresses terrorism, as authors make attempts to explain, but with some indeterminacy preserved, the terrorist's mind-set. For instance, Mohsin Hamid's *The Reluctant Fundamentalist* stages a former American's self-explanatory monologue about his newfound hatred of his country (modeled after Camus's *The Fall* [1956]).[26] The novel's narrator, as Medovoi has noted, undergoes the inverse of the traditional immigrant narrative: the protagonist and narrator Changez is a successful young Pakistani American who experiences an epiphany in which he rejects the values of the West and decides instead to join the terrorists.[27] Importantly, though, this decision is not entirely staged as a rational one. Instead, his decision is part of a traumatic disappearance of a girlfriend, a neurotic breakdown that both enables and doubles his decision to refuse the American Dream. The disappearance of his girlfriend and his epiphany about the corruption of his lifestyle turn Changez into an "incoherent and emotional madman, flying off into rages and sinking into depressions."[28] Such a process of doubling suggests that leaving the Americanized and Western community is an act that perhaps cannot be understood, or at least easily represented from within the perspective of that community, as fully rational. Hamid's portrayal, too, works within the paradigm of imagining terrorist consciousness as pathological, despite its frequently sympathetic portrayal of its narrator and his anti-American views.

In fact, the text that provides the best template for the type of sympathetic pathology that I explore in the final sections of this chapter comes from science fiction. As opposed to these particular examples of documentary and literary fiction, the Peabody Award–winning television show *Battlestar Galactica* seems in fact to encourage a more psychologically complex representation of fundamentalism and terrorism, despite imagining the fundamentalist terrorists in its plot as literal robots.[29] The show's opening miniseries depicts a terrorist attack that wipes out most of a multiplanet, space-going civilization. The eponymous "battlestar" warship is the center of a tiny remaining fleet of ships that remain after the attack, who are constantly besieged by the terrorist threat of their enemies, human-shaped robots called "Cylons." These Cylons take on human form and are visibly indistinguishable from people.[30] As such, they draw directly on the science fictional pod people tradition, from Heinlein, Finney, and Asimov to Dick and Stephenson, in the depiction of an undetectable threat walking in our midst. *Battlestar Galactica*'s most compelling inversion of this genre might be construed as borrowing in part from Ridley Scott's *Blade Runner,* that is, the hint that some of the robots might be more intensely human than the human colonists.

The show's chief adaptation for the post-9/11 moment, aside from the use of the Cylons as a destructive terrorist threat, is the incorporation of the robots' religious zeal. The human "colonists," as the show refers to them, are a classicized polytheistic group who come from planets named for the Roman zodiac signs; and their society bears many of the hallmarks of a tolerant multiculturalism, in which more religious characters coexist peacefully with secular characters. The Cylon robots, conversely, are monotheistic terrorists for whom peaceful coexistence is not an option, though they increasingly aim to convert the remaining humans to their religion. With the fictional civilization's classical bearings in planets like Sagittaron and Caprica, the religious allegory plays two ways: either the Cylons are intolerant Islamic terrorists, or, as the show progresses, they are increasingly the coming of Christ and Christianity that will save a postdiluvian remainder of the human race. The latter even comes complete with an immaculate—that is, human–Cylon—conception. The text's gritty realism and washed-out visual style, though, more strongly suggests the contemporary moment and, particularly in its first two seasons, the darker side of a coming fanaticism. Within that framework, too, *Battlestar Galactica* allegorizes many scenes and issues from the War on Terror, including searches for nuclear weapons, security measures aimed at wiping out terrorism, the suppression of civil liberties, the use of torture in interrogation scenes, and the sexual humiliation of prisoners.[31] Through these scenes and the occasional use of a Voigt–Kampff-like test to determine if a colonist is, wittingly or unwittingly, actually a Cylon agent, *Battlestar Galactica* also rehearses many of the ethical questions about the dehumanization of terrorists and cultural Others that other texts in the human programming tradition have asked. For instance, the protagonists eventually find themselves in a prison situation where they act as vigilante terrorists and flirt with the idea of carrying out suicide bombings themselves. The program also replays the sleeper-agent plot of *The Manchurian Candidate* with the young pilot Boomer, who eventually makes an assassination attempt on the commander of the fleet, Admiral Adama. Other episodes take shape from the supposed healing properties of Cylon blood to the friendliness and helpfulness of the Cylon Boomer, the scientist Baltar's erotic relationship with Caprica 6, and, at last, the revelations of the final five humans who are unwittingly Cylons.[32]

In part because *Battlestar Galactica* takes on the format of twenty-first-century long-running serial television, its four full seasons eventually subvert most of the generic expectations that accompany the human programming narrative, in the service of surprise endings and plot twists. Contemporary science fiction about selves and others tends overwhelmingly to emphasize the porousness of distinctions between us and them: that porousness also characterizes the post-2008 explosion of zombie narratives (in which a participatory costuming culture has arisen), and it stretches back to the development of "queer" alien narratives in the 1980s and 1990s, such as Octavia Butler's Lilith's Brood trilogy and the later entries in the Alien franchise.[33] *Battlestar Galactica* likewise does its most compelling work in undermining its own initial premise, that of a stable fight between a fragile but resilient and valorous humanity and the genocidal robotic terrorists who resemble jihadists. Unlike the theme-and-variations of classic science fiction television, in which new situations or enemies refresh a stable dynamic between human selves and alien others, each new variation or development in the show further destabilizes the barriers between the humans and the terrorist robots. But where in science fiction, the fictional device of the robotic body helps to hypostasize the abstract differences between a liberal democracy and a prophetic totalitarian fundamentalism, I will suggest that pathologies and diagnoses perform a similar function in literary fiction whose representations are confined to realism.

Falling Man and the "Ones Who Think Alike, Talk Alike"

Representing fundamentalist terrorists in realistic literary fiction brings a distinct set of challenges. A demand for seriousness, and for gravitas and accuracy, accompanies the attempt to represent terrorism directly. The success of creating a round terrorist character depends on a novelist's ability to inhabit a fundamentalist terrorist's point of view convincingly, a challenging task that often meets with failure. Take, for instance, Michiko Kakutani's *New York Times* review of John Updike's novel *Terrorist,* a review that draws on the language of automatism and human programming to point out the novel's representational shortcomings.[34] In the novel, Updike depicts a young boy named Ahmad who is swept into a radical Islamist plot in post-9/11 New Jersey, as well as the high school guidance counselor who convinces him, at the last possible second, not to carry through with the bombing. Although many readers and critics found *Terrorist* compelling, Kakutani judges Updike's efforts a failure. And she does so by calling Ahmad "a completely unbelievable individual," "more robot than human being," a "cliché" and a "one-dimensional stereotype" who has "been brainwashed to spout jihadist clichés." Through this description, Kakutani establishes a curious criterion for the novel of terror's success, something like a relatable roundness of character that might push such serious fiction beyond "cliché" and toward more original characterizations and insights. That roundness can be contrasted with a ready-to-hand "one-dimensional" "robot" characterization of a terrorist whose consciousness fails to stray outside the terrorist plot. Ironically, actual recent science fictional depictions of robot terrorists tend to be, like *Battlestar Galactica,* more nuanced than Kakutani's estimation of Updike's final novel.

DeLillo's *Falling Man* considers both terrorist consciousness and Islamophobia at once. This novel has been analyzed in dozens of academic articles through approaches focused on perception, affect, and, most prominently by numbers, mourning, trauma, and the ineffability of the 9/11 event.[35] For my part, I contend that examining the images of automatism in the text can reveal some of its most productive work in imagining the difficulties and limitations of conceiving a subhuman terrorist consciousness. The novel rehearses many of the public-sphere debates about 9/11 and terrorist consciousness, and one of its most interesting subplots involves the character Lianne's rejection of the explanations for terrorism that she finds around her. Lianne's search for a rounder representation of the terrorist begins with a consideration of the Koran, which Lianne's acquaintances are reading and "earnestly trying to learn something, [to] find something that might help them think more deeply into the question of Islam."[36] Lianne, despite her curiosity, finds herself skeptical of this move and wonders if this kind of learning amounts to a "determined action that floats into empty gesture."[37] She continues, "Maybe they were persisting. They were serious people perhaps. She knew two of them but not well. One, a doctor, recited the first line of the

Koran in his office: This Book is not to be doubted."[38] Particularly for the Western reader who is using the Koran as a sort of skeleton key to unlock the complexities and varieties of Islam, this single line's interdiction of doubt constructs the Muslim as a necessarily unthinking, slavish follower of dogma. From Lianne's perspective, the significance of the citation lies not just in its being the first line of the Koran but in its being the line that this "serious" doctor, looking to understand more about Islam, chooses to repeat as though it reveals something profound about the religion. The problem with this approach is neatly encapsulated in the following sentence: after the doctor's quotation from the Koran, the reader is told that Lianne "doubted things, she had her doubts."[39] The first use of "doubt" appeals to her own subjective experience of the world, because doubt forms a significant part of her religious disposition, as well as that more broadly of a modern secular and skeptical outlook on the world. How could she, a doubter, relate to or understand such a believer? But the second "doubt" in the sentence, in the plural, marks the doctor's interpretive dead end: it seems that this well-meaning doctor has only picked out the line that confirms a preexisting belief in the blind, will-less follower. The novel introduces skepticism regarding this endeavor, even though it would seem to mirror the project of the novel as a whole, that of engaging with the unfamiliar to better understand the terrorists. In this way, DeLillo acknowledges the limitations of his attempt to represent the terrorist, though the novel's terrorist, Hammad, experiences doubt about his role in the bombing he will carry out.

That doctor's reliance on the Koran can illuminate, too, the way that Updike's *Terrorist* represents Islamic consciousness as a short-circuit between text and consciousness. In Kakutani's pan of the "robotic" young terrorist, Updike's shortcoming is that of taking the same starting point, as a novelist, that *Falling Man*'s Koran-reading doctor does. Kakutani sees, in a similar fashion to DeLillo's Lianne, that only a short-circuit between holy text and religious consciousness could allow Ahmad to be characterized almost exclusively through the sacred words of Mohammed. In the novel, the boy Ahmad spends much of his time in the novel either repeating or contemplating the true meaning of various suras of the Koran. On the novel's first page, Ahmad states that all the teachers he has to listen to in the New Prospect, New Jersey, school system "lack true faith; they are not on the Straight Path; they are unclean."[40] By page six, he is, with his teacher, Shaikh Rashid, reciting and pondering a sura of the Koran about the "Crushing Fire" and deciding that his teacher is not a true enough believer.[41] The Koran serves as an anchor for Ahmad throughout the text; it colors his interpretations of the situations around him and almost stands in for the interiority of his character. This Koran-toting and sura-quoting constitute the literary technique that allows Updike to represent a young, troubled, and fatherless boy who would join, unquestioningly, a fundamentalist terrorist plot in post-9/11 New Jersey. Of course, not all readers see Ahmad as a "robot," and, moreover, literary characters are necessarily textual automatons, who embody concepts and social positions with varying degrees of additional detail.[42] Updike's Ahmad is pathologized insofar as his personal history of father-lessness and adolescent insecurity constitutes the bodily backing to his propensity for terror. Importantly for that novel, Ahmad's racial makeup—his half-Egyptian heritage—contributes to the story only insofar as it informs Ahmad's choice to follow Islam as a way of connecting to that absent father. That is, to the extent that Ahmad's characterization as a terrorist is bodily, it is pathological and not racial, hewing to the dogma line and not to the color line. As such, it's a "robotic" consciousness that Kakutani picks out of that novel (rather than a racist one) and that DeLillo's Lianne finds as the danger in others' impatient attempts to understand the terrorist.[43]

In a less guarded instance, however, Lianne, too, is momentarily seduced by this mode of thinking of Islam as a religion of slavish and blind obedience. When she contemplates the Middle Eastern music that leads to a breakdown of civil neighborly interaction, her rationalization rests on a similarly structured dehumanization: "They're"—Muslims, but a meaningfully vague "they"—"the ones who think alike, talk alike, eat the same food at the same time. She knew this wasn't true. Say the same prayers, word for word, in the same prayer stance, day and night, following the arc of sun and moon."[44] In this short, complete paragraph, DeLillo embeds an uneasy self-doubt in the middle sentence, ambiguously negated by the initial description's elaboration in the third sentence.

The notion of a determining discipline, of simultaneity and training that shape a deadened and uniform consciousness, is surely an incomplete explanation, and in this passage, neither the reader nor Lianne is permitted to take it at face value. The expected causality is reversed in these lines, as though Lianne first imagines a group of people who "think alike" and then casts about for material explanations in shared rituals that might amount, from a secular perspective, to conditioning. In the text, then, doubt becomes a way to avoid the short-circuit between a discourse and a consciousness that is produced directly from it, even as the text seems unable to produce a better alternative. This problem springs from imagining the stunted consciousness of the other as emerging fully formed out of a totalizing idea or discourse or form of discipline. "She knew this wasn't true." Lianne here realizes that such a way of imagining Muslims cannot be either ethical or correct and that the vision of the Muslim as automatic fanatic—that is, someone who wholly believes, and is overwhelmed by religious fervor and enthusiasm—must be flawed, even if she cannot articulate why. She seems to half-know that she is participating in the long tradition of linking Islam and fanaticism, of imagining the Muslim's "abstractive fanaticism" in Hegel's footsteps.[45] In Lianne's description, the blind believer springs wholly formed from the ritual repetition of the act of prayer, such an abstract and conformist being that he cannot, perhaps, be imagined as fully human. Her spiritual quest leads her to the philosophy of Kierkegaard and to the *Falling Man* performance art piece, in which the performer David Janiak jumps in imitation of a World Trade Center 9/11 suicide, to hang suspended, not unlike the human automaton, between personhood and objecthood.[46]

Falling Man's formal approach to the terrorist's consciousness develops through a comparison between Keith Neudecker's PTSD and the 9/11 hijacker Hammad's terrorist consciousness. The sections of *Falling Man* that deal with victimhood depart in several ways from the typical narrative of working through trauma, which often involve either performative acts of telling or writing or else the completion of a quest, such as that in Foer's *Extremely Loud and Incredibly Close* (2011).[47] In what seems on the surface a strange choice for *Falling Man* as a novel about 9/11, in the aftermath of the event, the 9/11 survivor and PTSD sufferer

Keith Neudecker becomes a compulsive poker player, taking long trips alone to weeks-long high-stakes poker tournaments. From his weekly game with friends before the attacks, Keith associates poker with a sense of extreme discipline, which DeLillo compares with the discipline of the "fanatic."[48] In those poker games, a series of arbitrary constraints regarding the game and the environment—only five-card stud, no food or drink but brown liquors, no cigarettes and compulsory cigars—cements the small group together, and they feel a comfort, every hand that is dealt, as they announce the game: "the words became a proud ritual . . . five card stud."[49] The poker game thus embodies in the novel a strange combination of discipline and compulsive repetition, absolute chance and the possibility of agency in responding to the chance: "the cards fell randomly, no assignable cause, but he remained the agent of free choice."[50]

The poker game is both a purification of the abstract principle of chance that governs his finance job in the World Trade Center and a regulation of chance itself: by playing five-card stud, the odds are not unnaturally enhanced by the addition of wild cards or extra cards, nor does the opportunity to discard give the player any choice about what he is dealt.[51] At one of the poker tournaments near the end of the novel, Keith wonders if he has become a "humanoid robot," a comparison that aptly names the cycle of compulsive avoidance that is a result of his traumatic experience.[52] As if to underscore the strength of this avoidance, it is soon after this moment that the narrator reveals the nightmares, symptoms of PTSD, that have plagued Keith through the course of the three years following the attacks: "These were the days after and now the years, a thousand heaving dreams, the trapped man, the fixed limbs, the dream of paralysis, the gasping man, the dream of asphyxiation, the dream of helplessness."[53] In light of this description about dreams of powerlessness and paralysis, the obsession with chance and agency in the poker game evokes both the common question of the survivor—why did I survive, and not the others?—and the sense of control that comes from repetition, as in Freud's famous "Fort-Da" game, in which repeating a small gesture or game enacts a symbolic form of control that mirrors an uncontrollable larger situation.

As Keith abandons his family to play high-stakes poker, he becomes less sympathetic as a character, and his compulsive behavior brings him into implicit comparison with the terrorist Hammad, whom we witness in the process of doubting and reconsidering his role in the terrorist act to come. As he continues his existence playing poker in high-class hotels, Keith imagines himself as "rigidly controllable . . . a robot dog with infrared sensors and a pause button, subject to seventy-five voice commands."[54] Hammad's consciousness is described according to a similar metaphor of a rigidly limited consciousness. He feels, as many critics of the novel have noted, "the magnetic effect of plot . . . [which] closed the world to the slenderest line of sight, where everything closed to a point"; this textual metaphor for Hammad's limited horizon of choices closely resembles Keith's at the poker table, as that of the mechanical robot, subject to programmable commands.[55] The process of converting toward an extreme form of consciousness is, as in the implied narratives of both brainwashing and Stockholm syndrome, often understood as a traumatic one, as a process in which rationality, individual personality, and freedom are supposedly erased to make way for a new, much more rigidly delineated cognitive map of the world. DeLillo's text, by juxtaposing post-traumatic and terrorist consciousnesses, challenges readers to consider whether these forms of consciousness must necessarily be considered, as they have so often been represented in popular media, as forms of subhumanity.[56]

Sympathetic Pathology in *Homeland*

The text from the past decade that most insistently compares terrorist consciousness and pathological consciousness has been Showtime's *Homeland*. Often compared to *The Manchurian Candidate* in reviews and even promotional materials, *Homeland* charts the return of a Sergeant Nicholas Brody (Damian Lewis) after eight years in al-Qaeda captivity. Carrie Mathison (Claire Danes) plays the CIA agent who has received the tip that al-Qaeda has "turned" a prisoner into a sleeper agent. Carrie, who suffers from bipolar disorder, then bears the burden of proving that Brody is the "turned" prisoner, acting as a sleeper agent terrorist. Borrowing heavily from the police procedural, Carrie disobeys orders and breaks the law with the best of them, in a search for the proof that will stop Brody and his supposed plot. Thus the stage is set for a complex and rich serial television program that toys with the Patriot Act, torture, terrorism, surveillance, government corruption, the covert sphere, the mandates of the twenty-four-hour news cycle, drone strikes, and fanaticism. As such, *Homeland* is a clear example of what Timothy Melley has called the "geopolitical melodrama," a genre defined by a constitutive tension between terrorist threats and the potential overreaches of the national security state.[57] Because *Homeland* is defined by this tension, it is perhaps also inevitably politically ambiguous. From a liberal perspective, it might mobilize dissent against Obama's continuation of the drone program, the prison at Guantánamo, and, in retrospect, the PRISM surveillance program.[58] From a conservative perspective, *Homeland* encourages a "whatever it takes" approach to fighting terrorism, and, like Kathryn Bigelow's *Zero Dark Thirty* (2012), it tends toward justifying the use of torture against terrorism suspects. As such, *Homeland* puts on display the overreaches of the national security state and the violence of the terrorists without firmly denouncing either. In the marketing environment of contemporary television, *Homeland*'s use of the geopolitical melodrama format allows it to take on complex and political hot-button topics without alienating either conservative or liberal viewer bases. Rather than its party politics, then, *Homeland*'s most significant aspect for my purposes is the role of popular psychology in its approach to representing fundamentalist or terrorist consciousness.

Homeland has been one of the more popular and critically acclaimed geopolitical melodramas, one that yokes together the human automatism I've explored throughout this book with questions about surveillance and the national security state. *Homeland* deals with the sleeper agent and the problem of terrorist consciousness, but this consciousness is not quite the mechanically triggered automatism of *The Manchurian Candidate* or its post-9/11 remake. Jonathan Demme's *Manchurian Candidate* remake, while it draws in some ways from the post-9/11 resurgence of "brainwashing" discourse I discuss in the introduction, draws corporations like Halliburton into its critical view, such that even the intimation of Islam is written out of the text entirely.[59] Though both *Homeland* and *The Manchurian Candidate* films do evoke PTSD or "combat exhaustion" as possible covers for more sinister problems, in *Homeland,* Islam itself stands as the marker of Brody's transformation into an enemy of the state. Moreover, *Homeland* describes the PTSD peculiar to the POW cell, and Carrie's spying reveals a man who, early on, finds comfort in crouching in the corners of rooms, wakes screaming from nightmares, and has difficulty connecting with others.

Though Brody does suffer visible symptoms of PTSD, *Homeland* also explores how little the visual field reveals about his psychology. In a version of Melley's geopolitical melodrama retooled for the contemporary surveillance state in particular, *Homeland* combines invasive visual surveillance of Brody with clear indications of that surveillance's impotence. In this significant strand of the first season, Carrie conducts in-home surveillance of Brody following her hunch that he is the "turned" American whom her informant has mentioned. Carrie's surveillance stakeout takes place primarily on her living room sofa, a clear nod to the relationship between the television viewer and the voyeur, and to the entertainment value Carrie gains from watching Brody's home life while eating out of takeout containers at home. Predictably, Carrie learns much more about the Brody's family's personal life and his wife's affair than she learns about his terrorist involvement, not least because the surveillance team has forgotten to bug the family's garage, the location to which he furtively and suspiciously retreats from time to time. Brody's true mental state and allegiances, that is, withhold themselves from view. The show's first episode borrows from the original Israeli version of the show *Prisoner of War* (which focuses on three returned POWs and has no Carrie figure or thriller plot) in giving Brody a finger-twitch signal. When Brody appears on television, Carrie notices, his fingers twitch in a strange way that she suspects may be a code. Evoking Gilbert Ryle's classic problem in philosophy of mind—how can we tell, with certainty, a twitch from a wink?—the show's use of the surveillance gaze leaves the CIA and the viewer trapped in Brody's exterior. This unusual finger twitch presents the same problem as the surveillance camera's capture of Patty Hearst's movements in the previous chapter: is it the volitional movement that signals collaboration with the enemy, or not? The CIA cannot break the finger-twitch code, if, in fact, it is a code at all.

Homeland's engagement with surveillance reaches its peak when it stages a pastiche of the Union Square scene from Francis Ford Coppola's *The Conversation* (1974), here set near Washington, D.C.'s, Dupont Circle.[60] The scene highlights Carrie's team's homophobic denunciation of a queer Saudi diplomat, their inability to distinguish Walker from another African American character, and their failure to protect American citizens.[61] As in *The Conversation*, the gaze of surveillance is simultaneously too invasive and completely impotent. *Homeland*'s focus on surveillance continually emphasizes what the observed body does not reveal, particularly when the aim of the surveillance is to determine whether Brody is a terrorist, a question that evolves into the more practical question of whether he will carry out a suicide-bombing plot. The viewer and Carrie grasp for criteria that might identify Brody as a terrorist, but the body gives away less than we expect. Instead, the show uses the complexity of ideas surrounding coercive persuasion and pathology to plumb those depths of Brody's consciousness that do not inhabit his moving image alone.

Homeland's use of third-person surveillance in the present is intercut with first-person flashbacks that investigate Brody's conversion to Islam as the next set of clues for plumbing deeper into his terrorist consciousness. We see the beginnings of these flashbacks, and then present-day Brody's act of slipping off to the garage to perform his prayers is framed as the suspenseful end of the show's second episode. The attentive viewer recalls at this point that when Brody's prison camp appeared on the news in the first episode, its location had been announced as "near Damascus," placing Brody in the footsteps of Saul of Tarsus's conversion. The show's emphasis on his conversion to Islam treads nearly in the Islamophobia that al-Arian and Shabi see in the show, and, like *Falling Man*, *Homeland* explores the Muslim-hating tendencies of largely sympathetic, otherwise liberal and enlightened characters. A child in Quaker school gives voice to the opinion that "all Arabs believe . . . [and] want the same thing, which is to annihilate us." Even beyond the mistake of assuming that all

Muslims are violently inclined fundamentalists, his slippage between Arabs and Muslims—"all Arabs believe"—is not an uncommon one in the contemporary United States, one that Volpp gestures toward when she describes the fuzzy "racialization" of Islam.[62] Later in the season, Carrie, too, finds her way around an investigative dead end by deciding that a colleague of hers must be a double agent: "Well, he *is* a Muslim," she says to herself and her companion, to convince us of his wrongdoing.

Even discounting the show's implication in these characters' Islamophobia, the seemingly unavoidable paradigm for representing Brody's conversion places it in terms of the coercive persuasion and the *topos* of the cult. When the conversion is presented as coerced, it equates Islam with both blind obedience and an inevitable betrayal of the United States, as though the audience, too, will anticipate Carrie's line: "Well, he *is* a Muslim." His meticulously orchestrated imprisonment near Damascus follows many of the criteria, discussed in the preceding chapter, that had been set forth by sociologists and psychologists of cult programming: the charismatic leadership, enclosed physical and social space, and specialized language that characterize the *topos* of the cult are all present in his captivity with Abu Nazir. And, perhaps more important for the viewer, these steps in his imprisonment and conversion follow the well-established generic conventions for the Stockholm syndrome narrative. After Brody undergoes much isolation, deprivation, and torture, Nazir appears for the first time to offer him food and drink. Nazir's manipulation is an obvious one to the audience, and the terrorist leader cheats his way into Brody's affections. As this manipulation continues, Brody's old allegiances are nonetheless broken down. He is made to participate in the torture of his fellow captive, Tom Walker, after which Nazir leads him to believe that he has killed his friend by beating him to death. This step in Brody's indoctrination conjures intense feelings of guilt and confusion, and he is vulnerable and bereft of connections to his country. Thus broken down, Brody appears in a key early scene in the show, his cell door left accidentally ajar, his path to escape clear. As the confused and bewildered Brody makes his way to the door outside, he stops to see all his captors praying in an adjacent room; he turns around, and he stays to pray. This conversion to Islam is framed in overexposed, hazy shots that signal a mystery, between grace and delirium, even as the conventional forms of coercive persuasion hang heavily in the atmosphere.

Despite the prevalence of this kind of imagery, the notion of "brainwashing" as such lurks at the margins of *Homeland*. In an interview during the course of the first season, the actor who plays Brody, Damian Lewis, fields a question about *The Manchurian Candidate* by saying, "In *The Manchurian Candidate* the Lawrence Harvey character is brainwashed. This is not that. Brody is a proactive force—he acts as a soldier, but for emotional and personal reasons rather than for faith-based reasons. So he chooses to act; he's not brainwashed into doing something."[63] Lewis's distinctions point again at the show's complex knot of "personal reasons," "faith-based reasons" and "brainwashed" reasons, and he places the latter two in an ambiguous relation to each other. This tangle aside, Lewis also seems careful to distinguish *Homeland* as something more than a rehash of a science fiction cliché. The dialogue of *Homeland,* moreover, largely avoids the term, though its place in the plot supports the notion that "brainwashing" stands in as a less sophisticated way of understanding Brody.

Pointedly, the term *brainwashing* is used to describe the second sleeper agent, Tom Walker, who had been in the same POW camp with Brody. When Walker surfaces at last in the United States with an assassination agenda, Carrie enlists his wife, Helen Walker, to help the CIA ensnare him. Helen, having been told that her husband was dead, tells Carrie, "Now you tell me that he is [alive] but they've turned him into some kind of monster. 'Brainwashed': that was the word you used. Planning an attack on his own country." "That's what we believe," replies Carrie. Though insufficient for Brody, the notion of "brainwashing" provides a suitable shorthand for Walker, a relatively flat character whose motivations are depicted more straightforwardly than Brody's. Walker's willing participation in an attack on the United States recalls the story of another Walker: John Walker Lindh, the "American Taliban" member who was captured in Afghanistan at the end of 2001.[64] In news reportage, in legal strategy, and particularly in interviews with shocked friends and families, John Walker Lindh was referred to repeatedly as having been "brainwashed." Brody utters the term "brainwashed" only in the course of denial, in the suicide tape that imagines that the media will call him "brainwashed." This avoidance of "brainwashing" allows Brody to lay claim to a rational thought process and to reasons for bombing his country that ought to be listened to. However, the show also positions "brainwashing" as the too-simple answer to the question of Brody's consciousness, one that it explores through a knot of reasons that draws together personal revenge, religious beliefs, coercive persuasion, and pathologies. The high culture of the War on Terror rejects the simple answer of "brainwashing," just as Kakutani rejects an insufficiently developed character as a "robot." These rejections of "brainwashing" and the "robot" share much in common: they reject the trace of a science fictional conceit that they deem inadequate to the serious task of representing the terrorist. That gesture of differentiation in turn marks the ideal investigations as high cultural products, capable of handling subtlety, ambiguity, and complexity.

Nevertheless, *Homeland* insistently restages Brody's conversions in much the way that *The Manchurian Candidate* and other films from the genre of human programming have done. One episode late in the first season narrates a more comfortable version of his manipulation, wherein he is taken from his jail cell, cleaned up, and assigned the mentorship of Abu Nazir's son Issa. This relationship with Issa, who is killed in a drone strike, constitutes one of Nazir's best manipulation strategies. As the story of Issa unfurls through flashbacks, in Brody's present, he is captured, placed in a cell resembling the house with Issa, and given a mysterious injection before videoconferencing with Nazir.[65] With Brody again in the closed space and without a clue about the substance with which he has been injected, Nazir uses his persuasive and manipulative power over Brody successfully. Even Nazir's insistence that Brody is acting freely reads with ironic resonances: "I pray every day that you never lose sight of what you committed to do in Issa's name. I never wanted or planned any of this . . . it was your choice doing what you are doing after what happened that day." Again, the text's complexity for the viewer comes in deciding to what extent Brody has made this "choice" freely.

Just as *Falling Man* uses indirect comparison in addition to pointing its gaze directly at the terrorist, *Homeland* makes use of a sympathetic pathologized character, Carrie Mathison. Carrie's maverick investigation strategy, of contriving to bump into Brody outside a Veterans Affairs group therapy meeting, also establishes a strong parallel between them. At the meeting, she strikes up a friendship with Brody based on their shared experience of combat zones in the Middle East. "Nobody understands anything" back home, she and Brody agree. From this establishment of an affinity through different degrees of PTSD, *Homeland* begins, in a manner even more sustained than *Falling Man,* an extended comparison between Brody's mental illness and Carrie's. In addition to having PTSD symptoms after serving in the Middle East, Carrie is bipolar, an illness she medicates with clonazepam and which she must hide from her superiors at the CIA for fear of being disqualified from service. In the manner of other recent crime dramas featuring flawed protagonists (e.g., HBO's *The Wire* [2002–8], the BBC's *Luther* [2010–], or Showtime's *Dexter* [2006–13]), Carrie's illness also makes claim to being the source of her brilliance, a sort of flawed spark. (The role fits well within Danes's career, which she has joked is a sort of tour through the *DSM*.)[66] When she goes off her medications, the walls of her home become, after a three-day flurry of activity, the chaotic collage of the crazed conspiracy theorist, though there is method, and a crucial clue, contained within it. Carrie's bipolar disorder plays out as a dogged persistence, an idée fixe that is further fueled by her pursuit of the dangerous pastime of having sex with Brody even as she becomes more certain he is a terrorist. *Homeland* diffuses Carrie's pathology between bipolar disorder, PTSD, *amour fou,* a self-destructive love of danger, and a kind of genius for doing her job. At the season's end, Carrie chooses to undergo ECT to forget Brody and to lead a normal, healthy life—a procedure that visibly registers a sort of self-imposed exorcism even as it evokes the well-known scene of ECT in Milos Forman's *One Flew over the Cuckoo's Nest.*

I want to suggest, too, that this affinity between Brody and Carrie goes beyond conventional forms of mutual respect or complementarity, wherein the cop and the robber need each other. Instead, their characters

Damian Lewis as Nicholas Brody and Claire Danes as Carrie Mathison in the first season of Showtime's *Homeland* (2011–). The series explicitly compares the damaged psychology of Brody, a victim of coercive persuasion and a PTSD sufferer, with that of Carrie, who suffers from PTSD and bipolar disorder.

become sympathetic and complex in equal measure. When the vengeful CIA director ejects Carrie from the force, confiscating her documents, on the pretense that her illness disqualifies her, the viewer roots against the managerial or bureaucratic gaze that discounts Carrie's ideas on the basis of her pathology. By the same token, then, we must ask if we similarly reduce or discount the ideas and positions of the terrorist Brody, which are suspended between the explanations of coercive persuasion and PTSD and righteous indignation at the U.S. government's actions. Moving from the crime drama's convention of the flawed investigator, *Homeland* works at expanding the viewer's horizon of sympathy from Carrie to Brody, asking us to acknowledge the dignity of one broken individual, then another. The same can be said of the parallel structure governing Keith and Hammad in DeLillo's *Falling Man*.

Still, like Carrie's ambiguously pathological spark, Brody's terrorist pathology and willingness to participate as a suicide bomber are distributed between several possible causes and motivations, split between rational dissent, PTSD, and his conversion to Islam. As with many of the other fictions of human programming we have seen, *Homeland* insistently repeats the scenes and conditions of programming and the actions Brody is programmed to carry out. Under what conditions of force, *Homeland* asks, does persuasion amount to coercive persuasion or programming? Under what conditions must we acknowledge the suicide bombing as a rational act? In the episode in which Brody picks up his explosive vest near the historic Civil War battlefield at Gettysburg, he describes the act of suicide bombing in terms of both American patriotism and grace within Islam. At the battlefield, he tells his children about Joshua Chamberlain's heroic charge in the battle, a stampede toward the approaching enemy using only bayonets: "it was so unexpected—it was so crazy—that the line was held that day. All because of a schoolteacher from Maine who was willing to do what was necessary for a cause that he believed in." Trying on his suicide vest moments later, Brody says, "I read that part of the blast goes upward . . . so that the head is cleanly severed, and normally found intact." And then, in un-subtitled Arabic, "La illaha illa Allah" ("There is no deity except Him," Koran 2:163). Such a juxtaposition echoes that between Carrie and Brody once again, between

the "crazy," self-destructive act we interpret as patriotic and the act of suicide bombing.

"Brainwashing" would be the easy solution, the simplest and perhaps
even vulgar answer to the question of the terrorist's state of mind. *Homeland,* in participating fully in the high culture of terrorism, traffics instead
in a complex moral and pathological ambiguity, and it perhaps most
fully takes up Jacqueline Rose's charge not to "dehumanize" the suicide
bomber in its decision not to name his acts as wholly pathological. His
daughter convinces him not to carry out the suicide bombing of the vice
president in the climax to the first season, such that he renounces both
the "absolute" mission of holy war *and* his vengeance in the name of
Abu Nazir's son Issa. By the second season of *Homeland,* Brody's motivations and allegiances become clearer, and his character is more fully
humanized. There, the supposedly "fundamentalist" ideology that Brody
adopts amounts primarily to questioning the U.S. government's actions.
As the second season of *Homeland* progresses, Brody develops as a character who thinks that drone strikes and governmental cover-ups are a
bad thing and who just happens to be a practicing Muslim.[67]

While much of the media coverage of terrorists produces them as
irrational and subhuman refusers of democracy, we see in literary, cinematic, and televisual narrative forms an overwhelmingly different drive.
Battlestar Galactica, Falling Man, and *Homeland* all point toward the commonalities between ordinary citizens and terrorists, between our ordinary
psychological makeups and those that we consider to be "extremist"
forms of consciousness. Where the accusation of "brainwashing"—or,
similarly, "radical Islam"—allows us to dismiss an ideology as unthinkable, the attempts to understand the terrorist embodied in these texts
seem a laudable, if perhaps unreachable, goal in the expansion of audiences' sympathetic horizons. They attempt to understand the terrorist
through affinity, in the sympathetic valence of the word "understand"
rather than the distant valence of "comprehension" or "reconnaissance"
implied by scenes of surveillance and expert study. Indeed, if the high
culture of the War on Terror has something to offer the contemporary
security state, perhaps it is the challenge of incorporating this sympathetic valence of "understanding" into observation and interpretation.

This paradigm of representation has been prominent enough to constitute a final development in the serious use of human automaton imagery in the United States, though in the conclusion, I turn to the science fiction of Daniel Suarez, which usefully reframes the question of contemporary individual agency in terms of network structures, of mediators and intermediaries, in a way that both draws on discourses of automatism and attempts to move more fully beyond them.

CONCLUSION

Automatism and Agency

O ver the course of the preceding chapters, we have seen the peculiarly American dialectic of self and other made visible by the circulation of the human automaton, which has described American freedom
by contrast with nondemocratic forms of unfreedom. Those descriptions
have been developed, and sometimes challenged, in cinema, literary fiction, science fiction, sociology, psychology, political science, and news
media. Here I can recast that historical narrative of *Human Programming* in two additional ways: first in terms of its scientific iterations, then
through its formal literary and cinematic dimensions. While the chapters of *Human Programming* have recounted in a roughly chronological
sequence the various uses of human automatons in public discourse,
the book has also endeavored to describe the various horizons of scientific possibility through which human automatism has been imagined.
In those terms, the sequence of the chapters also maps onto scientific
paradigms through which external techniques might, in those horizons
of possibility, penetrate the innermost regions of the mind or alter one's
beliefs. These begin with the advent of cinematic propaganda and behaviorist conditioning (which could occur with the aid of psychotropic
drugs that began in the 1950s), all of which made totalitarian automatons
and communist brainwashing imaginable in the 1940s and 1950s. Next
came an emphasis on institutions and social spaces, where the human
automaton was used in texts that highlighted the conditions of its social

production, in both progressives' visions of society-as-total-institution and the sociology of cults. The metaphors of computer programming were largely confined to science fiction and hyperbole (such as Friedan's and Hochschild's robotic women), though cybernetic approaches to the mind as a computer-like system reverberated through sociology and psychology in the later decades of the century. In a final development, the representations of cults and terrorist cells since the 1970s have increasingly imagined trauma and mental pathologies such as post-traumatic stress disorder as modes through which individuals might be unintentionally but nonetheless radically altered.

And as I discussed in the introduction, the cultural connotations of these scientific paradigms have been defined through their uses in a variety of areas. As a form of what Michael Rogin has called "demonization," the human automaton is characterized by the mind's unnatural separation from the body and its lack of adaptability.[1] As I discussed in the introduction, atavistic and eugenic paradigms of racial thinking early in the twentieth century imagined the reversion of particular racial groups toward those of animals, in a hereditary hierarchy of races. While eugenics had imagined the mind's reversion to animal instinct, as though the mind were being absorbed by its bodily, animal nature, the human automaton was consistently imagined as unnatural; with the human automaton, the mind seems as though it has been separated from the body, or some force aside from the mind drives movements or speech. This paradigm of representing individuals as mindless or programmed was particularly adaptable to antiracist public discourse in the wake of World War II and to the United States's enemies in the axis powers, communism, and totalitarianism. That antiracism and independence of race-based forms of exclusion then aided the circulation of human automaton imagery of unfree subjects between the 1970s cult scares and the early 2000s War on Terror.

What has emerged in an admittedly piecemeal fashion over this historical narrative is also a typology of the human automaton as a major literary figure of the twentieth century: (1) a source of comedic and uncanny effects that (2) works across cinematic and prose representation to (3) figure forth the various possible disconnections between the body's visible actions and the mind's invisible intentions. First, from the range of more and less disconcerting confusions between people and objects come humorous and uncanny effects of automatic movement and speech. These effects, such as Captain Marco's rote repetition of a programmed stock phrase in *The Manchurian Candidate*, play on our understanding of the presence of thought within speech and of agency within action. These purely formal effects, as I recounted in the introduction, were exploited in theater, dance, philosophy, and fiction before the twentieth century, but the media technologies of sound film, with their formally intrinsic play on the presence and absence of their performers, frequently took advantage of the aesthetic effects of human automatism. The cinematic effects of automatism have often functioned through qualities of motion, of deadness in the eyes (as with the factory workers in *White Zombie* or Raymond Shaw in *The Manchurian Candidate*), and of repetitions, while textual adaptations of the image often described such qualities on the page. In terms of their critical ambitions, literary and cinematic displays of automatism fall along a spectrum between (1) placing these images on gratuitous display (as when Neal Stephenson's *Snow Crash* devises automatons as an effect of science fictional "cool") and (2) deploying the confusion between personhood and objecthood in an ethically charged fashion, drawing attention to readers' criteria for counting members of their community (as with Ralph Ellison's use of literary automatons to criticize sociologists' refusal to acknowledge African Americans as fully human). Of course, the critical ambitions of texts can, in the absence of archival evidence, be ultimately undecidable, as with *The Manchurian Candidate*, which positions the viewer both as the ethically motivated Marco who wants to save the brainwashed Shaw and as the Russian brass who continually want to watch him kill.

As I mentioned in the introduction, Jacques Rancière's work in *Disagreement* helps me to couch the ethical stakes of this political acknowledgment within our historical moment. The acknowledgment of subjects in Rancière's political philosophy is explicitly the "counting" of an individual's or group's voice. The decision of whether to count an individual's or group's voice is often a matter of perception or aesthetics, a question that literary and cinematic human automatons often raise formally. The process of politics, and of democratization, for Rancière, is that of expanding the count of the whole and acknowledging a wider range of individuals as having perspectives worth hearing and understanding. What my investigations have emphasized, however, is that strategies for reducing the count of the whole continue to proliferate and that, with increasing prominence since the second half of the twentieth century, those who wish to reduce that count often appeal to forms of scientific knowledge to justify themselves. In the 1950s and 1960s, the discourse of "brainwashing" sought to exclude the voices of communists in an aggressive fashion. In another historical coexistence of inclusion and exclusion, the moment of 1990s multicultural acknowledgment was also marked by discourses about new religious movements and fundamentalism that would carry into the War on Terror. If exclusions on the scientific basis of biological race have been on the decline since World War II, then discourses that purport to exclude or to speak for other subjects on the basis of sociological or psychological expertise have multiplied in the era that saw brainwashing, "cult" discourse, expertise on terrorism, and representations of fundamentalism as a form of unfree consciousness. Although we have been on guard against discourses where we have excluded subjects on the basis of their racialized bodies, *Human Programming* suggests that we be vigilant about describing, or speaking for, subjects as though they have subhuman, environmentally determined (in the strongest sense), or automatic minds.

And this ethical focus might in turn be applied to literary criticism and cultural theory by way of *Human Programming*'s observations about institutions in chapter 2. In the progressive uses of automaton imagery I examined there, I identified what I called an anti-institutional ethos in the postwar United States, and I traced how, in works by Ken Kesey, Ralph Ellison, and Betty Friedan, that anti-institutional ethos derives from antitotalitarian imagery. By beginning to trace a genealogy of that anti-institutional ethos, I hoped to emphasize its historical contingency within both cultural production and academic criticism and theory. Scholars who make use of Louis Althusser's and Michel Foucault's influential accounts of institutional power might easily assume that the institution is automatically the *opposite* of individual empowerment. In an unusual circuit, the institution becomes the thing we critique, more or less self-consciously, in the name of a kind of individual agency (and against forms of human automatism) that have been borrowed, perhaps unwittingly, from the grammar of American individualism and automatism in the second half of the twentieth century. Mark McGurl's study of the institution of the MFA program and Michael Szalay's study of the literary culture of the Democratic Party suggest one kind of start to this scholarly project: to describe the ways that institutions shape action and agency, without assuming that the form of every institution is the panopticon. As Lily Kay puts it, the form of the institution "enables and constrains" the actions of its participants, and those enabling factors can only emerge when we imagine institutions as something other than a zero-sum game of individual freedom in tension with institutional coercion.[2] (David Foster Wallace's final two novels also gesture in this direction, for instance, as they find complex individuals within highly controlled institutions—the tennis academy, the halfway house, and the offices of the Internal Revenue Service—in which even the most vulnerable subjects shaped by those institutions still take part in rich individual and collective forms of experience.)

As a final brief examination of human automaton imagery, I conclude here by taking up Daniel Suarez's 2006 novel *Daemon,* which places familiar images of humans-become-automatons in conversation with an alternative scheme for imagining agency in the contemporary United States. In *Daemon,* Sobol, a computer programmer with a terminal illness sets up a networked computer virus that can read the news and access enormous amounts of money. The computer programmer has set up an automated, weaponized home, and he has also laid plans for financial schemes and extortion rackets that computer programs will carry out after the event of his death. After the programmer dies and the news of his death appears online, the Daemon contacts the recently fired journalist Anji Anderson with a robocall, offering exclusive information on the story that will help to jump-start her career. Like the recent American television political thriller *House of Cards* (2013–), *Daemon* emphasizes the symbiotic relationship between the journalist and her information source, which remains mutually beneficial even when one partner is a machine. It becomes clear as Anji continues reporting the Daemon's news—which in turn spurs on other events triggered in the Daemon's plan—that the agency of the partnership itself lies in the feedback loop through which the journalist disseminates information that both advances the journalist's career and causes events the Daemon desires. The journalist Anji is an agent in the sense of being a mediator of information (filtering it through new sources), but to no greater extent than the computer program (which finds and processes information after Sobol's death) is an agent in this system. Another major agent of the Daemon, Brian Gragg, accomplishes tasks for the Daemon in a smooth continuum with the world of a computer game that Sobol had designed. These are scenarios of machine agency in which none of the participants is exactly free or unfree on an individual basis but all spur events into motion.

These scenarios make good examples of what the sociologist Bruno Latour has written about as "actor networks."[3] Whereas Latour's eye has been trained on seeing a wider variety of *nonhuman* actors, this networked understanding of agency can also be usefully applied to human subjects. The characters in *Daemon* share in an actor network with a computer program, through which the agency of the humans and computers involved is considered on an equal footing. Latour's primary distinction within actor networks is that between "mediators" and "intermediaries": actors that actively reshape the information or entities they transmit and actors that passively transmit information or entities they pass along.[4] By writing news stories or carrying out tasks in the real world, these characters are just such mediators. Such a view of their actions is one that is diametrically opposed to the often all-or-nothing view of the human automaton characters we have seen throughout *Human Programming*, particularly in fictions wherein the total unfreedom of an automaton props up the apparently sublime total freedom and agency of a protagonist character, as in the action-genre dynamics of Stephenson's *Snow Crash*.

Daemon goes on to explicitly compare the characters' networked agency to a "human automaton" version of agency. In another subplot, the novel demonstrates a case of the Daemon's total control over an individual by rehearsing a scenario of human programming much like those we have seen through the course of the present book. Charles Mosely, an African American convict in his thirties, is working at an in-prison phone bank when the Daemon calls him and offers him freedom. It orchestrates a prisoner transfer to a new facility whose computer records indicate that he is due to be released the following day. He is brought into the organization by being drugged and forced to watch a two-day-long interactive film that tests his psychological parameters, a situation that echoes the scenes of conditioning of *A Clockwork Orange* (Stanley Kubrick, 1971) and that, more recently, have been reprised in the Dharma Initiative of the television show *LOST* (2004–10). The Daemon replaces the total institution of the prison with a system of reward and punishment, ranging from large sums of money to the threats of life imprisonment and death, and a computer network that can always find him. Mosely is then employed as a hit man, and the news report after his first hit reveals him to be one among thousands of such troops for the Daemon. The Daemon effectively harnesses these human agents as its puppets through simple systems of reward and punishment, though when it threatens to kill those who do not cooperate, the agents seem more like slaves than networked associates. Most important, though, *Daemon* treats the forms of power that the reasonably realistic computer program has and the power that individuals have within the same plane, that of the actor network. These human agents are passive intermediaries or active mediators in the network in much the same way that different aspects of the Daemon's computer code are.

At the end of *Daemon*, Suarez rehearses the stock scenario in which computers take control of the world and eliminate human agency. As the setup to the novel's sequel, *Freedom*, the Daemon tells the protagonist, Sebeck, "I suspect that democracy is not viable in a technologically advanced society. Free people wield too much ability to destroy. . . . If you fail to prove the viability of democracy in man's future, then humans will serve society—not the other way around."[5] The Daemon network is thus posed as yet another technology—like behaviorist conditioning or programming the brain like a computer—that poses a threat to individual agency and autonomy, such that it becomes the opposite of democracy. Accordingly, the sequel rehearses many questions about the importance of individual freedom. But the structure of the Daemon program itself is instructive and points in a different direction altogether. In deriving its power essentially from network effects—from being centrally situated in networks that give it access to information and cash flows—the Daemon also suggests that we can reframe our conception of agency around the analysis of such network effects. This dimension of Suarez's novel thus suggests that we shift our gaze away from the invisible mental agency of the individual body and toward networks of connection if we want to understand individuals' power within and across the societies in which they are connected.

Daemon's focus on the networked nature of agency—that our networks define the reach of our ideas and our actions—helps to reveal a final paradox inherent in the human automaton as a literary figure. When we fix our gaze on the individual body as the source of action, the Kantian antinomy on which the automaton image is based—is this body the author of its actions, or are its actions unfree?—remains perpetually insoluble. Individual agency retains a sort of aesthetic mystery when we veil it in the surface of the individual body, its motions and its speech acts. Daemon, and Latour's actor-network theory, suggest that the way around this paradox, as with Xeno's paradox of the arrow that never leaves the room, is to shift our frame of reference. When we examine not the precise origin but rather the spread of ideas and actions, we can make visible a version of individual agency that depends on our networks, on the reach of our ideas and the echoes of our actions.[6] It is this dimension of connectedness that constitutes the major blind spot of the gaze that investigates the individual body's automatisms, in much the same way that connections between prisoners are written out of Jeremy Bentham's ideal panopticon.

In terms of the stakes and methodology of literary study, likewise, my goal throughout Human Programming has been to treat the agency of literature in a similar, networked mode. That is, I have held back from ascribing a wholly autonomous power of refusal, critique, or separation from scientific or capitalistic forces in literary depictions of human automatons, despite the compelling ethical suggestions many of these creative works have made. While I can echo the ethical imperative to acknowledge others as humans rather than imagining them as subhuman, my task has been primarily one of observing and tracing. Latour suggests that critique has "run out of steam," but that doesn't mean that literary criticism must give up on literature's political impact.[7] Instead, we need to be more creative about how we measure the political impact of literature, without attributing to it the sublime agency of its autonomous separation from the world, and likewise assuming that the sciences, bureaucracies, or the like are machinelike forces of total repression. (To do so would be to make a mistake that's structurally identical to those of Edward Hunter.) Certainly, scientific discourses have had the normalizing function that Foucauldian scholarship has emphasized, but perhaps this ought not to be our first assumption when approaching literature and science, so that we don't unwittingly replicate C. P. Snow's "two cultures" divide—we stand too much to gain by seeing the ways they work in tandem. Beyond a paradigm in which literature "critiques" scientific discourses, there are surely new modalities of literature and film's political impact waiting to be explored. The human automaton's impact has been most profound in the ways that it has influenced the political language of American culture. The best this book's observations can offer is to suggest the extent and saturation of that figure in U.S. culture by sketching how it has passed in and out of disparate realms of discourse, including social science, literature, film, political theory, and mass media. As such, the political agency of literary texts themselves has come through their complex feedback loops with other forms of discourse. And this seems like a first step in imagining our way beyond a depiction of agency that has consistently depended on the ability to envision its opposite, the human automaton.

Notes

Introduction

1. George W. Bush, "Address to a Joint Session of Congress and the American People," September 20, 2001, http://georgewbush-whitehouse.archives .gov/.

2. Don Oldenburg, "Stressed to Kill: The Defense of Brainwashing," *Washington Post*, November 21, 2003, C1. The occasion for this story about the resurgence of "brainwashing" was the Patty Hearst–esque legal defense offered by the Washington, D.C., sniper suspect Lee Boyd Malvo, later convicted. On Lindh's whiteness as a factor in this media coverage, see Sean S. Brayton, "An American Werewolf in Kabul: John Walker Lindh, the Construction of 'Race,' and the Return to Whiteness," *International Journal of Media and Cultural Politics* 2, no. 2 (2006): 167–82.

3. Sharon L. Crenson and Martha Mendoza, "Mind-Control Defense Won't Wash, Some Say," *Los Angeles Times*, April 7, 2002, A21.

4. Eric Bailey, "Response to Terror: Shy Youth's Action Mystifies Family, Friends," *Los Angeles Times*, December 3, 2001, A4.

5. See Oldenburg, "Stressed to Kill," and Abd Samad Moussaoui and Florence Bouquillat, *Zacarias My Brother: The Making of a Terrorist* (New York: Seven Stories Press, 2003).

6. Scott Shane, "From Minnesota to ISIS," *New York Times*, March 22, 2015, A1.

7. George F. Kennan, "Totalitarianism in the Modern World," in *Totalitarianism*, ed. Carl J. Friedrich (New York: Grosset and Dunlap, 1964), 19.

8. Ibid., 19–20.

9. Edward Hunter, *Brainwashing: The Story of the Men Who Defied It* (New York: Farrar, Strauss, and Cudahy, 1956), 25.

10. Jack Hitt, "The Year in Ideas: The Return of the Brainwashing Defense," *New York Times Magazine,* December 15, 2002, supplement, 6.116.

11. Minsoo Kang, *Sublime Dreams of Living Machines: The Automaton in the European Imagination* (Cambridge, Mass.: Harvard University Press, 2011).

12. Karel Čapek, *R.U.R. (Rossum's Universal Robots),* trans. Paul Selver and Nigel Playfair (New York: Samuel French, 1923).

13. Such undecidability bespeaks a certain exhaustion of symptomatic reading strategies, which Katrina Mann illustrates in arguing against several symptomatic readings and for another, in "'You're Next!' Postwar Hegemony Besieged in *Invasion of the Body Snatchers,*" *Cinema Journal* 44, no. 1 (2004): 49–68. Of *Body Snatchers,* she writes, "The film's invasion discourse was less specifically concerned with bureaucrats, autocrats, Reds, and radiation than with the potential disruptions of the gender, racial, and sexual status quo such phenomena threatened to bring about" (49)—the first being accounts of previous symptomatic readings—and ultimately she argues that it is more concerned with "racial 'amalgamation'" in African American mobility (54), and she also associates the aliens with "Mexican migrant laborers" (56). Although we can admit that viewers might associate the film's content with any of these contexts, I argue in what follows that they cannot be definitive measures of a text's political impact.

14. Kang, *Sublime Dreams of Living Machines*; Lisa Zunshine, *Strange Concepts and the Stories They Make Possible* (Baltimore: Johns Hopkins University Press, 2008); Bill Brown, "Reification, Reanimation, and the American Uncanny," *Critical Inquiry* 32, no. 2 (2006): 175–207; Despina Kakoudaki, *Anatomy of a Robot: Literature, Cinema, and the Cultural Work of Artificial People* (New Brunswick, N.J.: Rutgers University Press, 2014).

15. See Masahiro Mori, *The Buddha in the Robot: A Robot Engineer's Thoughts on Science and Religion,* trans. Charles S. Terry (Ann Arbor, Mich.: Kosei, 1981).

16. Donna Haraway, "A Cyborg Manifesto: Science, Technology, and Socialist-Feminism in the Late Twentieth Century," 1985, in *Simians, Cyborgs, and Women: The Reinvention of Nature* (London: Free Association Books, 1996).

17. Cary Wolfe, in *What Is Posthumanism?* (Minneapolis: University of Minnesota Press, 2009), associates "liberal humanism" with "autonomy" (xv). N. Katherine Hayles, in *How We Became Posthuman* (Durham, N.C.: Duke University Press, 1999), associates "liberal humanism" with "self-possession" (4). (Haraway's "A Cyborg Manifesto" does not expand on humanism.) Because neither author fully elaborates on the tradition of liberal humanism, it's possible to see the field as vulnerable to the misconception that humanists (from Kant and Hegel to Arendt and Sartre, say) believe naively in the self-transparency and unimpeachable willfulness of the human subject. For my part, I see a productive tension between the academic projects of humanism and posthumanism, and particularly in popular expressions of humanist self-presence, transhumanist utopianism, and the dystopias inhabited by technologically posthuman automatons.

18. Wolfe does note that "posthumanism . . . comes before and after humanism," but he situates its "post" in terms of a philosophical approach to the subject. Wolfe, *What Is Posthumanism?*, xv.

19. Shu-mei Shih, "Is the *Post-* in Postsocialism the *Post-* in Posthumanism?," *Social Text* 30, no. 1 (2012): 27–50. Working in a slightly different vein to my own, Shih arrives at a striking conclusion about posthumanism: "When certain people have not been considered and treated as humans, posthumanism serves as an alibi for further denial of humanity to these same people. Cybernetics might be a step beyond old-fashioned Enlightenment humanism, technologically speaking, but the newly emerging subjects of history—colonized peoples, women, minorities of all kinds—need to be respected and dignified as humans first. Here the question is not about temporality—the subhumans are asking for old-fashioned humanism and hence are hopelessly anachronistic—but about priority within the same historical moment shared and lived by all" (30). See also a similar argument, as well as a deployment of automaton imagery in a consideration of Third World feminism, in Rey Chow, "Postmodern Automatons," in *Feminists Theorize the Political,* ed. Judith Butler and Joan Wallach Scott (New York: Routledge, 1992), 111.

20. William James, *The Principles of Psychology* (New York: Dover Thrift Editions, 1990), 92.

21. See D. M. Wegner, *The Illusion of Conscious Will* (New York: Bradford Books, 2003).

22. William James, *Talks to Teachers on Psychology: And to Students on Some of Life's Ideals* (Rockville, Md.: Arc Manor Press, 2008), 43.

23. James, *Principles of Psychology,* 116.

24. In *Principles of Psychology,* James responds to Huxley's automaton theory by elaborating the notion of consciousness as a "selective agency" (139). That selecting agency "sinks to a minimum" during "rapid, automatic, habitual action" (139). And "the operation of free effort, if it existed, could only be to hold some one ideal object, or part of an object, a little longer or a little more intensely before the mind" (579). The most consequential of these selective agencies operates through our creation of whole situations even as we perceive only parts of them: we construct "objective reality par excellence" at any moment on the basis both of "present sensations" and of "absent ones," and it is the absent sensations that we mark as "significant" that often matter the most, say, when we imagine a whole society at the moment we vote for a political candidate (139). In this way, James locates agency not at all in the body and its habits of motion but wholly in the mental actions of perception and selection. See also Sara Ahmed's *Willful Subjects* (Durham, N.C.: Duke University Press, 2014) for an alternate genealogy of will and of "willfulness," which Ahmed imagines as running counter to institutions' ordinary operations.

25. Immanuel Kant, *Critique of Pure Reason,* trans. Paul Guyer and Allen Wood (New York: Cambridge University Press, 1999), 115.

26. James, *Principles of Psychology*, 104.

27. In *The Cybernetic Brain: Sketches of Another Future* (Chicago: University of Chicago Press, 2011), Andrew Pickering postulates a "black-box ontology" that guides cybernetics and other twentieth-century scientific endeavors such as behaviorism (20). This image of the black box has its origins in an engineering problem, wherein an engineer must fix a "secret and sealed bombsight that is not working properly," thus operating something without regard to its inner workings (20). Such a black-box ontology evinces a Jamesian pragmatism that, in Pickering's account, privileges a new kind of experimentalism and creativity in American science.

28. Kerry W. Buckley, *Mechanical Man: John Broadus Watson and the Beginnings of Behaviorism* (New York: Guilford Press, 1989), 121–22. According to Buckley, this experiment was an explicit response to Freud's "Little Hans" case study, which involves a young boy's fear of a white horse: Watson set out to prove that reactions, such as those associated with fear, had everything to do with conditioning and nothing to do with the unconscious (in which he also did not believe).

29. See, e.g., Ellen Herman, *The Romance of American Psychology: Political Culture in the Age of Experts* (Berkeley: University of California Press, 1996).

30. Buckley, *Mechanical Man*, 101.

31. Kara Reilly's *Automata and Mimesis on the Stage of Theater History* (New York: Palgrave, 2011) discusses automatons in theater and dance in general terms and with reference to techniques used within dance and theater. Vsevelod Meyerhold's biomechanical theories of bodily movement in dance, in which emotion springs from movement, rather than vice versa, are discussed in Felicia McCarren, *Dancing Machines: Choreographies in the Age of Mechanical Reproduction* (Stanford, Calif.: Stanford University Press, 2003). *Petrushka* and *Coppélia* feature dolls animated by magicians. *The Triadic Ballet* features characters in highly geometrical and abstract costumes whose movement is constricted and clearly mechanical. Dance, an art form in which the movement of the body is also a primary element, also features the programmatic nature of bodily repetition, which these works exploit thematically.

32. Reilly, *Automata and Mimesis*, 241.

33. Kang's recent study *Sublime Dreams of Living Machines* offers a comprehensive history of the automaton in the Western imagination, from hydraulic moving fountains in ancient Greece through the European heyday of automatons in the seventeenth through nineteenth centuries and contemporary robotics. See also Jessica Wolfe, *Humanism, Machinery, and Renaissance Literature* (New York: Cambridge University Press, 2004).

34. See Julia Douthwaite and Daniel Richter, "The Frankenstein of the French Revolution: Nogaret's Automaton Tale of 1790," *European Romantic Review* 20, no. 3 (2009): 381–411. This recent study of revolution-era automatons in French and English literature contends that, "as a metaphor in philosophy and literature of the period, the term automaton *(automate)* was

invariably pejorative and synonymous with puppet or monkey: an empty-headed creature of habit" (392). Deidre Lynch points out one of the more significant instances of this pejorative metaphor in English literature, in Frances Burney's 1796 novel *Camilla, or, A Picture of Youth*, which likens the training of a young girl in the manners of society to an artificial set of constraints on personality that bears comparison with an automaton (192–98). Of course, such a metaphor is immediately ripe for satire and as such has political significance as well. Diderot's *La Religieuse* (1796), for instance, linked automatism and convent life, and in 1791, the Marquis de Condorcet wrote a fiction in which a clockmaker fashioned a royal family out of clockwork to satirical ends: because the mechanical king would have a limited range of motion, citizens could "declare him inviolable without injustice, and infallible without absurdity" (qtd. in Douthwaite and Richter, "Frankenstein of the French Revolution," 392).

35. Douthwaite and Richter, "Frankenstein of the French Revolution," 381.

36. T. S. Eliot, "The Waste Land," in *The Waste Land and Other Poems* (New York: Harvest, 1955), 256. See also "Convictions (Curtain Raiser)" and "Humouresque" in T. S. Eliot, *Inventions of the March Hare: Poems, 1909–1917*, ed. Christopher Ricks (New York: Harcourt Brace, 1996), 11, 325.

37. Mark Seltzer, *Bodies and Machines* (New York: Routledge, 1992), 76, 100. In a trajectory that this book as a whole traces, we see a parallel concern embedded in narrative that Timothy Melley, in *Empire of Conspiracy: The Culture of Paranoia in Postwar America* (Ithaca, N.Y.: Cornell University Press, 2000), 49, discerns as a form of "agency panic" that, mutatis mutandis, is widespread within postmodern American literature.

38. Michael Benson's *Vintage Science Fiction Films, 1896–1949* (New York: McFarland, 1985) lists many trick films involving supposed magnetism or automatic function of objects, along with a number of supposed automatic people and animals, American unless noted otherwise: *The Automatic Monkey* (1909, French), *The Automatic Motorist* (1911, British), *The Automatic Servant* (1908), *The Electric Policeman* (1909, French), *The Electric Girl* (1914), *The Mechanical Legs* (1908, French), *The Mechanical Husband* (1910), *The Mechanical Man* (1915), *Mechanical Mary Anne* (1910, British), and *The Mechanical Statue* (1907). See Tom Gunning, "Now You See It, Now You Don't: The Temporality of the Cinema of Attractions," *Velvet Light Trap* 32 (Fall 1993): 11, for the classic analysis of this period in early cinema.

39. See Noël Burch, *Life to Those Shadows* (Berkeley: University of California Press, 1990), for the distinction between early, often nonnarrative "primitive" representation and the increasingly codified "institutional" mode of Hollywood's continuity editing. See also Linda Williams, "Film Bodies: Gender, Genre, and Excess," *Film Quarterly* 44, no. 4 (1991): 2–13, and Miriam Hansen, "The Mass Production of the Senses: Classical Cinema as Vernacular Modernism," *Modernism/Modernity* 6, no. 2 (1999): 59–77. My comparisons

between a modernist problematic of agency and the aesthetics of cinematic motion are meant as a continuation of Hansen's titular suggestion.

40. In *Un chien andalou*, directed by Luis Buñuel (1928; Paris: Les Grands Films Classiques, 1987), there is a long segment with a woman who holds a severed hand and who stands motionless in the street in seeming imitation of it. Surrealist cinema, dance, comedy, and other stage and screen arts frequently establish mechanical-seeming parameters of motion beyond which they expand as narratives develop.

41. Michel Foucault, *Discipline and Punish: The Birth of the Prison*, trans. Alan Sheridan (New York: Vintage, 1977).

42. Ibid., 136, 141.

43. Ibid., 135.

44. See Giorgio Agamben, *Homo Sacer: Sovereign Power and Bare Life*, trans. Daniel Heller-Roazen (Stanford, Calif.: Stanford University Press, 1998), and Roberto Esposito, "Totalitarianism or Biopolitics? Concerning a Philosophical Interpretation of the Twentieth Century," *Critical Inquiry* 34, no. 4 (2008): 633–44. Agamben notices that Foucault, "in striking . . . never dwelt on the exemplary places of modern biopolitics: the concentration camp and the structure of the great totalitarian states of the twentieth century" (10). Agamben's *homo sacer* figure, as the sacrificial subject of biopolitics, holds some commonalities with the human automaton, but in the historical period that is my own focus, the mechanics and fear of human programmability are often foregrounded much more than the sacrificial or biological aspects of *homo sacer*'s "bare life." For his part, Esposito contrasts the historiography of the anti-totalitarian tradition with that of the biopolitical tradition.

45. Gilles Deleuze and Félix Guattari, *Anti-Oedipus*, trans. Robert Hurley, Mark Seem, and Helen R. Lane (Minneapolis: University of Minnesota Press, 1983).

46. Michel Foucault, preface to ibid., xiii, emphasis original. Although a thorough consideration of the French post-1968 context is beyond the scope of the book, both Foucault and Deleuze, in the texts mentioned, work to describe the disciplinary apparatuses of the postwar system in terms that imagine totalitarianism as an exaggeration of the unfreedom imposed by capitalist and institutional life.

47. Jacques Rancière, *Disagreement: Politics and Philosophy*, trans. Julie Rose (Minneapolis: University of Minnesota Press, 2004), 26.

48. Ibid.

49. Henri Bergson, *Laughter: An Essay on the Understanding of the Comic* (New York: Macmillan, 1911).

50. Ibid., 92. There has been much discussion of Bergson relative to the cinema, particularly with regard to the illustrative section of *Creative Evolution* that explains "The Cinematographical Conception of Time" (296–324). Anson Rabinbach discusses at length Bergson's conceptual debts to Marey's

photographic work in *The Human Motor: Energy, Fatigue, and the Origins of Modernity* (New York: Basic Books, 1990), 110–14. See also Deleuze's defense of the fluidity of time within film and other cinematic adaptations of Bergsonian concepts in *Cinema I: The Movement-Image,* trans. Hugh Tomlinson (University of Minnesota Press, 1987), 1–12. I take the cinematic instances of laughter without much regard for the medium's inherent spatialization of time.

51. Tom Gunning has indicated that he is "increasingly suspicious" of Bergson's total theory of laughter—Gunning develops a compelling theory of the "gag," "machines that produce nothing other than a process that destroys [themselves] . . . crazy machines" in "Mechanisms of Laughter: The Devices of Slapstick," in *Slapstick Comedy,* ed. Tom Paulus and Rob King (New York: Routledge, 2010), 138. The gag often involves the destruction (or self-destruction) of a machine that "normally embodies the instrumental logic of human behavior": it is "misuse" or "the machine turned against itself"—or its owner—that constitutes a gag (140). Gunning's focus on the machines in our surroundings offers a nicely comprehensive account of the slapstick gag, but I would insist on pointing back to the actor, who so often carries through the gag by rigidly expecting the machine to work in its ordinary way rather than figuring out, with the audience, that, for example, a string has been attached to the doorknob or that a certain umbrella is, beyond a doubt, broken. Gunning ultimately reads against Bergson's grain in asserting that the "mechanization of the human . . . recalls less Bergson's mockery of the inflexibility and absurdity of the human body, and more the way the mechanical endows the human with a sense of grace, perfection, and even freedom" (148). I do not read such a sense of "mockery" in Bergson's "Laughter." The inflexibility of the Tramp in *Modern Times,* for instance, is precisely only temporary, giving way to grace in other instances, a reading that I believe Bergson would readily affirm. See also Gunning, "Buster Keaton or the Work of Comedy in the Age of Mechanical Reproduction," in *Hollywood Comedians: The Film Reader,* ed. Frank Krutnik (New York: Routledge, 2003), 73–77.

52. Bergson, *Laughter,* 156.

53. Ibid., 145.

54. Minsoo Kang asserts that the power of the object is often the deciding factor of whether an automaton is sublime, humorous, or uncanny in *Sublime Dreams,* 44. In chapter 2, I argue that the dimension of community can be equally important, as I analyze several scenes of automatism in Ralph Ellison's *Invisible Man* (New York: Vintage, 1952) that are interpreted by various characters as humorous and uncanny at the same time.

55. Freud disagrees: "the motif of the seemingly animate doll Olympia is by no means the only one responsible for the incomparably uncanny effect of the story, or even the one to which it is principally due." Bill Brown recently reasserted the value of Jentsch's approach in an article on uncanny objects,

and together with him, I think it is reasonable to assess much of Freud's dismissal of Jentsch as a rhetorical gesture. Brown, "Reification, Reanimation, and the American Uncanny," *Critical Inquiry* 32, no. 2 (2006): 175.

56. Sigmund Freud, *The Uncanny*, trans. David McLintock (1919; repr., New York: Penguin Books, 2003), 143.

57. Michel Foucault, *The Birth of the Clinic: An Archaeology of Medical Perception*, trans. Alan Sheridan (New York: Vintage Books, 1994), 29.

58. For instance, Bill Brown, "Reification, Reanimation, and the American Uncanny," locates one of these when he asserts that the uncanny quality of automaton dolls that represent African Americans springs from the American history of slavery: "the apprehension that within things we will discover the human precisely because our history is one in which humans were reduced to things (however incomplete that reduction)" (207). In his reading of "The Sandman," Freud explores how disavowed and repressed desires invest this scene of ontological uncertainty with a specific, additional urgency, which Jentsch, looking at the doll alone, cannot explain. What Freud adds to Jentsch's analysis, then, even in his disagreement, is that these scenes of ontological uncertainty are consequential, rather than merely curious, because human automatism can be invested with a variety of erotic attachments and social meanings.

59. Kant, *Critique of Pure Reason*, 115. This conflict is also the subject of Kant's third antinomy in the first *Critique* (484–89), and the supposition of freedom's a priori existence is a key foundation for his ethics in the *Grundwerk* and the second *Critique*. Immanuel Kant, *Practical Philosophy*, trans. Paul Guyer and Allen Wood (New York: Cambridge University Press, 1999), 94–96, 139–40. Mutatis mutandis, this problem could also be said to appear in the choice between the knowable totality and the unknowable infinity of the subject in Emmanuel Levinas's philosophy, and in the "calculable" being of substance as opposed to the comparatively withdrawn being of *Dasein* in Martin Heidegger's work.

60. On philosophical zombies, see Robert Kirk and Roger Squires, "Zombies v. Materialists," *Proceedings of the Aristotelian Society, Supplementary Volumes* 48 (1974): 135–63, for an early articulation of the question, and for a longer discussion of subsequent arguments, see David J. Chalmers, *The Conscious Mind: In Search of a Fundamental Theory* (New York: Oxford University Press, 1997), 93–208, and Robert Kirk, "Zombies," in *The Stanford Encyclopedia of Philosophy*, Summer 2012 ed., ed. Edward N. Zalta, http://plato.stanford.edu/.

61. On Boas's significance in this regard, see Kristina Klein, *Cold War Orientalism* (Berkeley: University of California Press, 2003), 11, and Mark Greif, *The Age of the Crisis of Man* (Princeton, N.J.: Princeton University Press, 2015), 42.

62. While the eugenics movement had entailed more than just justifications for racism, as Daylanne English has shown, "since [the mid-1940s] to

label any sort of social policy or theory 'eugenic' has been effectively, and usually rightly, to malign it." English, *Unnatural Selections: Eugenics in American Modernism and the Harlem Renaissance* (Chapel Hill: University of North Carolina Press, 2004), 177.

63. Nikhil Pal Singh, *Black Is a Country: Race and the Unfinished Struggle for Democracy* (Cambridge, Mass.: Harvard University Press, 2005), 8. As Klein, *Cold War Orientalism*, 40, has likewise noted, "Secretary of State Dean Acheson, who described the poor state of domestic race relations as the 'Achilles' heel' of U.S. foreign relations, warned that discrimination caused significant 'damage' to America's world standing and 'jeopardize[d]' the strength of its alliances. A source of 'constant embarrassment' abroad, it created a 'formidable obstacle' to the achievement of Washington's goals."

64. Karl Gunnar Myrdal, *An American Dilemma: The Negro Problem and Modern Democracy* (New York: Pantheon, 1944).

65. Colleen Lye, *America's Asia: Racial Form and American Literature, 1893–1945* (Princeton, N.J.: Princeton University Press, 2005), 56.

66. Cultural representations of the Occupation have been discussed in Tony Williams, "White Zombie, Haitian Horror," *Jump Cut* 28 (1983): 18–20; in Jennifer Fay, "Dead Subjectivity: *White Zombie,* Black Baghdad," *CR: New Centennial Review* 8, no. 1 (2008): 81–101; in Rowe's discussion of Hurston; and in Mary A. Renda, *Taking Haiti: Military Occupation and the Culture of U.S. Imperialism, 1915–1940* (Chapel Hill: University of North Carolina Press, 2000). See also Susan Zieger, "The Case of William Seabrook: *Documents,* Haiti, and the Working Dead," *Modernism/Modernity* 19, no. 4 (2012): 737–54.

67. William Seabrook, *The Magic Island* (New York: Literary Guild of America, 1929), 101.

68. Ibid.

69. Ibid.

70. Ibid.

71. Ibid.

72. Eugene O'Neill, *The Emperor Jones* (Cincinnati, Ohio: Stewart Kidd, 1921), 5, 51.

73. Robert Heinlein would use automaton and zombie interchangeably for parasitic aliens as late as 1951 in *The Puppet Masters* (1951; repr., New York: Baen, 2010), 22, 58.

74. In what we might say is a transitional moment away from nationalist biological essentialism, the narrator describes the German love of regimentation as an explanatory framework: "The German people have an inborn natural love of regimentation and harsh discipline. [Hitler] could give them that."

75. Martin J. Manning and Herbert Romerstein, *Historical Dictionary of American Propaganda* (Westport, Conn.: Greenwood Press), 84.

76. Hermann Rauschning, *The Revolution of Nihilism: Warning to the West,* trans. E. W. Dickes (New York: Alliance, 1939), 34. Quoted in Greif, *Age of the Crisis of Man,* 341.

77. Rauschning, *Revolution of Nihilism,* 34.

78. Kant, *Practical Philosophy,* 94–96, 139–40.

79. Franklin Delano Roosevelt, "Message to Congress, 1941 [Four Freedoms]," Franklin D. Roosevelt Presidential Library and Museum, http://www
.fdrlibrary.marist.edu/.

80. Friedrich Hayek, *The Road to Serfdom* (1944; repr., New York: Routledge, 2001), 104.

81. David Horowitz and Robert Spencer, *Islamophobia: Thought Crime of the Totalitarian Future* (New York: David Horowitz Freedom Center, 2011). As Horowitz and Spencer's title and publication company name suggest, the rhetoric of freedom and unfreedom, of combating a *1984*-esque totalitarianism, is alive, well, and extraordinarily flexible today (liberal pieties and the censure of racist speech being, from this perspective, totalitarian and freedom restricting). These books on the right also signal the extent to which *Islamophobia* has become a central, and contestable, term in these debates in the past several years.

82. I thus set aside the theological iterations of the question of free will in Protestant and particularly Calvinist thought, which constitute a significant lineage for discussions of the mind's relationship to the body in U.S. culture.

83. Donald Pease, "Exceptionalism," in *Keywords for American Cultural Studies,* ed. Bruce Burgett and Glenn Hendler (New York: New York University Press, 2007), 109. Pease expands on these points in *The New American Exceptionalism* (Minneapolis: University of Minnesota Press, 2009), where he writes, "U.S. citizens could express their belief that America was exceptional even though they harbored very different accounts of what that belief meant" (9), a point that dovetails with my account of the flexibility of freedom as an ideal.

84. Pease, "Exceptionalism," 109.

85. I'm thinking here of work such as Priscilla Wald, *Contagious: Cultures, Carriers, and the Outbreak Narrative* (Durham, N.C.: Duke University Press, 2008); Russ Castronovo, *Beautiful Democracy: Aesthetics and Anarchy in a Global Era* (Chicago: University of Chicago Press, 2007); Mary Esteve, *The Aesthetics and Politics of the Crowd in American Literature* (New York: Cambridge University Press, 2003); and Lauren Berlant, *The Queen of America Goes to Washington City: Essays on Sex and Citizenship* (Durham, N.C.: Duke University Press, 1997), while also working alongside the emphasis on archival and institutional analyses that characterize the Post-45 school of American literary studies.

86. In this regard, Wald's *Contagious* works along similar lines, in that it describes the ways in which a wholly different narrative trope—the outbreak narrative and "patient zero"—has circulated between disparate scientific discourses, film, and literature.

87. Ellison, *Invisible Man.*

88. Neal Stephenson, *Snow Crash* (New York: Spectra, 1992).

89. Don DeLillo, *Falling Man: A Novel* (New York: Scriber, 2007).

90. Daniel Suarez, *Daemon* (2006; repr., New York: Signet, 2009).

1. Uniquely American Symptoms

1. *The Manchurian Candidate*, dir. John Frankenheimer (1962; Beverly Hills, Calif.: MGM/UA Home Entertainment, 2004), DVD.

2. Joost Meerloo, *The Rape of the Mind: The Psychology of Thought Control, Menticide, and Brainwashing* (Cleveland, Ohio: World, 1956), 51.

3. Michel Foucault, *The Order of Things: An Archaeology of the Human Sciences,* trans. Alan Sheridan (New York: Vintage, 1970).

4. Ibid., 387. Timothy Melley also notes that the brainwashing discourse pits these models of the subject against one another in *The Covert Sphere: Secrecy, Fiction, and the National Security State* (Ithaca, N.Y.: Cornell University Press, 2013), 71.

5. For complementary historical accounts that place brainwashing in the center of U.S. Cold War history, see Susan Lisa Carruthers, *Cold War Captives: Imprisonment, Escape, and Brainwashing* (Berkeley: University of California Press, 2009), and Matthew W. Dunne, *A Cold War State of Mind: Brainwashing and Postwar American Society* (Amherst: University of Massachusetts Press, 2009).

6. William Seabrook, introduction to *The White King of La Gonave,* by Faustin E. Wirkus and Taney Dudley (New York: Doubleday, Doran, 1931), xii.

7. Theodor Adorno and Max Horkheimer, *Dialectic of Enlightenment* (1944; repr., London: Verso, 1979), x. During his exile in the United States, Adorno became the chair of a committee of sociologists working on *The Authoritarian Personality* (1950), centered on the question of whether fascism could possibly take hold in America. Their answer to that question was a rather disconcerting yes, an answer whose consequences I address in the next chapter, where I turn to the U.S. counterculture of the 1950s and 1960s. Although Adorno is the best known among scholarly audiences today through his aesthetic theory and analysis of the culture industry, other thinkers from his wave of intellectual immigration were more influential in framing the problem of totalitarianism for U.S. culture at large. See also the similar conclusions reached by Siegfried Kracauer in "The Mass Ornament," in *The Mass Ornament: Weimar Essays,* trans. Thomas Y. Levin, 75–88 (Cambridge, Mass.: Harvard University Press, 1995), and in *From Caligari to Hitler: A Psychological History of the German Film,* ed. Leonardo Quaresima (Princeton, N.J.: Princeton University Press, 2004).

8. Erich Fromm, *Escape from Freedom,* 2nd ed. (New York: Henry Holt, 1965); *Fear of Freedom* in the United Kingdom.

9. Ibid., 19. He cites John Dewey, *Freedom and Culture* (New York: Putnam, 1939), 44.

10. Fromm, *Escape from Freedom*, 232. In this treatment of fascism as a psychological, Fromm's influence on the Adorno-led *Authoritarian Personality* can be seen; Ellen Herman identifies both in turn as influential texts in the period in *The Romance of American Psychology: Political Culture in the Age of Experts* (Berkeley: California University Press, 1995), 182.

11. Fromm, *Escape from Freedom*, 209.

12. Ibid., 213.

13. Ibid., 223.

14. Ibid., 281.

15. Ibid., 283.

16. Ibid., 276.

17. Mark Wollaeger, *Modernism, Media, and Propaganda: British Narrative from 1900 to 1945* (Princeton, N.J.: Princeton University Press, 2006), 6. See also Edward Bernays, *Propaganda* (New York: Liveright, 1928); Bernays, Sigmund Freud's nephew, would later found public relations as a field. Priscilla Wald relates the histories of propaganda and public relations to the texts of the brainwashing episode in "The 'Hidden Tyrant': Propaganda, Brainwashing, and Psycho-politics in the Cold War Period," in *The Oxford Handbook of Propaganda Studies*, 109–30 (New York: Oxford University Press, 2013).

18. Meerloo, *Rape of the Mind*, 210, 211.

19. Ibid., 211.

20. Friedrich Kittler, *Gramophone, Film, Typewriter*, trans. Geoffrey Winthrop-Young and Michael Wutz (Stanford, Calif.: Stanford University Press, 1999), 149; Stefan Andriopoulos, *Possessed: Hypnotic Crimes, Corporate Fiction, and the Invention of Cinema* (Chicago: University of Chicago Press, 2008), 107; Alan Nadel, "Cold War Television and the Technology of Brainwashing," in *American Cold War Culture*, ed. Douglas Field, 146–63 (Edinburgh: Edinburgh University Press, 2005).

21. Michael Shelden, *Orwell: The Authorized Biography* (New York: Perennial, 1992), 470.

22. Versions of this notion of linguistic relativity (that language shapes thought) have also developed across the history of the discipline of linguistics, beginning in the eighteenth century and best known as the Sapir–Whorf hypothesis beginning in the 1920s. See John Joseph Gumperz and Stephen C. Levinson, *Rethinking Linguistic Relativity* (Cambridge: Cambridge University Press, 1996). Orwell's notion, as I explain later, is distinguished by focusing on the propagandistic shortening of words.

23. George Orwell, *1984* (New York: Signet Classics, 1950), 253.

24. Ibid.

25. Relative to Shklovsky and Orwell, Michael Clune, "Orwell and the Obvious," *Representations* 107, no. 1 (2009): 30–55, offers a counterintuitive reading of perception in Oceania, reading Winston as a typical totalitarian subject for whom "everything is seen as if for the first time, day after day, forever,"

such that the totalitarian state occupies the role of the artwork for Shklovsky (32). For my part, I understand the speakers of "duckspeak," discussed later, as the typical subjects of Oceania and Winston's character and perceptions as primarily a narrative device for showing Oceania's details to the reader.

26. George Orwell, "Politics and the English Language," in *The Orwell Reader: Fiction, Essays, and Reportage*, 355–78 (New York: Mariner, 1961).

27. Ibid., 355.

28. Orwell, *1984*, 48.

29. Ibid.

30. Hannah Arendt, *Eichmann in Jerusalem: A Report on the Banality of Evil* (1963; repr., New York: Penguin Classics, 2006), 47.

31. Ayn Rand, *Anthem* (1938; repr., New York: Signet, 1946).

32. In the discipline of linguistics, the behaviorist model, described later, would be wholly displaced soon after B. F. Skinner's *Verbal Behavior* (New York: Copley, 1957) was savaged in a famous review by a young Noam Chomsky. See Noam Chomsky, "A Review of B. F. Skinner's *Verbal Behavior*," *Language* 35, no. 1 (1959): 26–58. Chomsky's subsequent development of generative grammar presupposes a subject of language who addresses her semantic needs creatively and dynamically with whatever language is available.

33. John B. Watson, *Ways of Behaviorism* (New York: Harper, 1928), 81. Watson's theories of language warrant comparison, too, to those of J. L. Austin and the later Ludwig Wittgenstein.

34. Orwell, *1984*, 251.

35. Ibid., 150.

36. Ibid., 267, 189.

37. Ibid., 138.

38. I have found 577 published uses of the word *brainwash* and 1,927 of *brainwashing* in the 1950s, about half of which appear to be serious uses of the term (and not, e.g., ironic applications of the term as a disparaging remark about a domestic group). A sampling includes news articles and editorials such as "Brainwashing Methods," *Science News-Letter* 70, no. 14 (1953): 214; "More about Brainwashing," *Washington Post*, July 3, 1953, 10; "Expert Tells Severity of Brainwashing: No Man Can Endure It, Schwable Quiz Told," *Chicago Daily Tribune*, March 10, 1954, 19; "Antidotes for Brainwashing," *Los Angeles Times*, June 3, 1955, A4; "Scientist, Senator on Brainwashing," *Los Angeles Times*, May 5, 1956, A4; and academic work such as Meerloo, *Rape of the Mind*, and James G. Miller, "Brainwashing: Present and Future," *Journal of Social Issues* 13, no. 3 (1957): 48–55. See also Seed's discussion of Daniel V. Gallery, "We Can Baffle the Brainwashers," *Saturday Evening Post* 227, no. 30 (1955): 20, in *Brainwashing: The Fictions of Mind Control* (Kent, Ohio: Kent State University Press, 2004), 83. Most news reports on brainwashing consult scientific experts who confirm to some extent that specialized techniques can be used to

persuade captives; some also report politicians' statements on brainwashing, which are usually hortatory in the mode of Hunter's writings, discussed later.

39. Timothy Melley, "Brain Warfare: The Covert Sphere, Terrorism, and the Legacy of the Cold War," *Grey Room* 45 (Fall 2011): 18–41, 28.

40. U.S. Congress, House, Committee on Un-American Activities, "Communist Psychological Warfare (Brainwashing) Consultation with Edward Hunter" (Washington, D.C.: U.S. Government Printing Office, 1958), 6.

41. Hunter, "Communist Psychological Warfare," 6.

42. Edward Hunter, *Brain-washing in Red China* (New York: Vanguard Press, 1953).

43. Hunter, "Communist Psychological Warfare," 6.

44. Seed, *Brainwashing,* 81–82. See also Dunne, *A Cold War State of Mind,* 13–51, for a discussion of the twenty-one soldiers, and Charles W. Young, *Name, Rank, and Serial Number: Exploiting Korean War POWs at Home and Abroad* (New York: Oxford University Press, 2014).

45. Herman, *Romance of American Psychology.* Catherine Lutz also makes this case, with reference to Herman's work, in "Epistemology of the Bunker: The Brainwashed and Other New Subjects of Permanent War," in *Inventing the Psychological: Toward a Cultural History of Emotional Life in America,* ed. Joel Pfister and Nancy Schnog, 245–67 (New Haven, Conn.: Yale University Press, 1997).

46. Hunter, *Brain-washing in Red China,* 91.

47. Ibid., 11.

48. Orwell, *1984,* 210.

49. Hunter, *Brainwashing.*

50. Ibid., 12.

51. Hunter, "Communist Psychological Warfare," 25.

52. Hunter, *Brainwashing,* 25.

53. Hunter, "Communist Psychological Warfare," 15.

54. The virus had been discovered in the early 1950s, and the circulation of virus accounts in the media, as an unseen threat, often explicitly borrowed from anticommunist language, and vice versa, as discussed in Wald, *Contagious,* 157–212.

55. Hunter, "Communist Psychological Warfare," 10.

56. Hunter, *Brain-washing,* 17.

57. Hunter, "Communist Psychological Warfare," 2.

58. Hunter, *Brainwashing,* 25.

59. Ibid., 92.

60. Ibid., 93.

61. Ibid., 94. Also of importance in this discussion is the fact that the Korean War was the first major conflict in which the army was racially integrated. The dialogue in Hunter's text bears comparison with the first war film about Korea, Samuel Fuller's *The Steel Helmet* (1951; New York: Criterion Collection, 2007),

DVD, which features this racial tension prominently. In the film, a lost platoon captures a Chinese communist officer, and the communist attempts to stir unrest within the platoon by citing racial injustices. "You will have to ride on the back of the bus," he tells the African American doctor in the platoon, after which he goes on to torment the Japanese American soldier with the history of World War II Japanese internment in California. The film, which celebrates the bravery of the infantryman, is widely sympathetic to the American soldiers who face racism at home and does a less-than-convincing job of neutralizing the communist's critique. The Japanese American cites his patriotism in the European Theater of World War II, offering only that, "if we get pushed around at home, that's our business."

62. See Abbott Gleason, *Totalitarianism: The Inner History of the Cold War* (New York: Oxford University Press, 1997), 92–94.

63. Hannah Arendt, *The Origins of Totalitarianism*, 2nd ed. (New York: Harcourt, Brace, and World, 1958). The emphasis on totalitarianism as "scientific" is also present, but not so succinctly expressed, in the first, 1951 edition of Arendt's work, particularly where she discusses the prominent role of science and scientism in both Soviet and Nazi propaganda (345–50).

64. Ibid., 462.

65. Hunter, "Communist Psychological Warfare," 3.

66. Meerloo, *Rape of the Mind*, 82, 107.

67. Ibid., 109.

68. Ibid., 106, 136–39.

69. Vance Packard, *Hidden Persuaders* (New York: Pocket Books, 1958).

70. McLuhan, *Understanding Media: The Extensions of Man* (New York: McGraw-Hill, 1964).

71. Philip Deane, *I Was a Captive in Korea* (New York: W. W. Norton, 1953).

72. Qtd. in McLuhan, *Understanding Media*, 208.

73. Ibid., 209.

74. Gallery, "We Can Baffle the Brainwashers," 20.

75. Isaiah Berlin, "Two Concepts of Liberty," in *Liberty: Incorporating Four Essays on Liberty* (New York: Oxford University Press, 2002), 184.

76. Ibid.

77. Ibid.

78. As Steven Belletto has pointed out, state planning as a phenomenon was also a source of anxiety in the period, as evidenced by a conference on "Determinism and Freedom in Modern Science," in *No Accident, Comrade: Chance and Design in Cold War American Narratives* (New York: Oxford University Press, 2012), 20.

79. Heinlein, *Puppet Masters*.

80. Ibid., 133.

81. Ibid., 196, 339.

82. Ibid., 205; Leerom Medovoi, "The Race War Within," in *American Literature and Culture in an Age of Cold War: A Critical Reassessment,* ed. Steven Belletto and Daniel Grausam (Iowa City: University of Iowa Press, 2012), 170.

83. Paul Edwards, *A Guide to Films on the Korean War* (New York: Greenwood Press, 1997), 10.

84. Ibid., 10–11.

85. See, e.g., Matthew Frye Jacobsen and Gaspar Gonzalez's book that offers a pedagogical lens on the Cold War period, *What Have They Built You to Do? The Manchurian Candidate and Cold War America* (Minneapolis: University of Minnesota Press, 2006).

86. Tony Jackson, "*The Manchurian Candidate* and the Gender of the Cold War," *Literature/Film Quarterly* 28, no. 1 (2000): 34–40; Seed, *Brainwashing;* Michael Rogin, *Ronald Reagan the Movie: And Other Episodes in Political Demonology* (Berkeley: University of California Press, 1988). Michael Szalay provocatively connects the text to the Democratic Party's political culture through of the 1960s through its interest in the Beats and Buddhism, in "The White Oriental," *Modern Language Quarterly* 67, no. 3 (2006): 363–96. Szalay also reads the "hip" racial dynamics of the film in his *Hip Figures: A Literary History of the Democratic Party* (Palo Alto, Calif.: Stanford University Press, 2012), wherein Frank Sinatra promotes an image of a hipster John F. Kennedy against an outmoded McCarthyite parody of the Republican Party. See also Wald, "Hidden Tyrant," which reads *The Manchurian Candidate* relative to propaganda and public relations.

87. Susan Carruthers astutely mentions the importance of media manipulation in the film, such that another hidden purpose for the plot would be to make use of a television broadcast, in *Cold War Captives* (Berkeley: University of California Press, 2009), 223.

88. Jonathan Demme's 2004 remake of the film, *The Manchurian Candidate* (Hollywood: Paramount Home Entertainment, 2004), DVD, does a good deal to patch up these inconsistencies—Raymond Shaw becomes the candidate in question, Bennett Marco the assassin, and the mother a party official capable of winning Shaw the vice presidential nomination. In the original film, they claim that, as a Medal of Honor winner, Shaw will be "beyond suspicion," yet it would seem unnecessary to then disguise Shaw as a priest for the assassination if the Medal of Honor were to serve that purpose.

89. For histories of the terminology, see J. M. W. Binneveld, *From Shell Shock to Combat Stress: A Comparative History of Military Psychiatry* (Amsterdam: Amsterdam University Press, 1997), and Ruth Leys, *Trauma: A Genealogy* (Chicago: University of Chicago Press, 2000). *Post-traumatic stress disorder* was coined in 1981; both historians agree about the continuity between shell shock, combat stress, and PTSD. Gulf War syndrome, to which the 2004 remake of *The Manchurian Candidate* refers instead, is actually a different phenomenon, although many soldiers from that war also suffered from PTSD. In

Gulf War syndrome, U.S. military victims of friendly fire began having physical symptoms that were attributed to depleted-uranium-tipped bullets used in the First Gulf War. For more on the latter, see Bernard M. Rosof and Lyla M. Hernandez, eds., *Gulf War Veterans: Treating Symptoms and Syndromes* (Washington, D.C.: National Academy Press, 2003).

90. The *DSM–IV–TR* lists six diagnostic criteria for PTSD, namely, (1) a stressor (the event itself); (2) intrusive recollection (including "recurrent and distressing dreams of the event"); (3) avoidant/numbing (including "efforts to avoid thoughts, feelings, or conversations . . . activities, places, or people that arouse recollections of the trauma" and "inability to recall an important aspect of the trauma"); (4) hyperarousal (including "difficulty falling or staying asleep" and "irritability or outbursts of anger"); (5) duration of longer than a month; and (6) functional significance, that is, causing "distress or impairment in social, occupational, or other important areas of functioning." American Psychiatric Association, *Diagnostic and Statistical Manual of Mental Disorders*, 4th ed., text rev. (Washington, D.C.: American Psychiatric Association, 2000), 463.

91. The coincidence between Hunter's account of "Brainwashing and the Negro" and the nonwhite soldiers' resistance in *The Manchurian Candidate* is also noteworthy, particularly with regard to Marco's role as a sort of hipster and as an Italian American. He gets deliveries of interesting books from a "book store in San Francisco," which could well be Lawrence Ferlinghetti's City Lights—and as recently as 1949, Antonio Benedetto had changed his name to Tony Bennett to gain appeal with white, non–Italian American audiences. Condon, in *The Manchurian Candidate* (1958; repr., New York: Orion, 2013), emphasizes Marco's "all-American-ness" in a peculiar way: "He looked like an Aztec crossed with an Eskimo, which was a fairly common western American type because the Aztec troops had drifted down from Siberia quite a long time before the Spaniards of Pizarro and Cortez had drifted north out of the Andes and Vera Cruz" (30).

92. Michael Herr, *Dispatches* (New York: Knopf, 1977).

93. Seed, *Brainwashing*, 57.

94. The Kantian ethical subject, as discussed in the introduction, presupposes a noumenal or undetectable "freedom" to make ethical choices; the film's vision of the behaviorist subject is, by contrast, the sum total of the determinations or conditioning placed on the subject.

95. Wald also compellingly reads the appropriation of the Lincoln image through Iselin's public relations mastery in "Hidden Tyrant," 122.

2. Anti-institutional Automatons

1. Ken Kesey, *One Flew over the Cuckoo's Nest* (New York: Signet, 1963), 40, 59.

2. Ibid., 16.

3. Ibid., 278.

4. Alan Nadel, *Containment Culture: American Narratives, Postmodernism, and the Atomic Age* (Durham, N.C.: Duke University Press, 1995). While *Containment Culture* is ultimately framed around symptomatic readings over a longer historical duration, Leerom Medovoi also productively challenges the narrative implied by Nadel's title along similar lines to mine in "The Race War Within." Carruthers's *Cold War Captives* also helps to reframe a historical narrative of the Cold War around images of captivity abroad and at home, which link Korean War brainwashing images to work by Betty Friedan and others, as discussed later.

5. Mark McGurl, *The Program Era: Postwar Fiction and the Rise of Creative Writing* (Cambridge, Mass.: Harvard University Press, 2009), 206. I agree with McGurl's assertion here that the novel is one of the "classic anti-institutional texts of all time," an assertion that takes into account the irony of the novel's own institutionalization within high school and college syllabi, which has paradoxically "helped reel untold thousands of unsuspecting 'disaffected youth' back into the educational groove" (202).

6. On organizations as open systems, see W. Richard Scott, *Organizations: Rational, Natural, and Open Systems*, 4th ed. (Upper Saddle River, N.J.: Prentice Hall, 1998), 21; McGurl, *Program Era*, 192–97; and the brief discussion of Erving Goffman in the following pages. Scott notes that, "with very few exceptions—such as inmates in 'total institutions,' for example, concentration camps or cloisters (Goffman, 1961)—participants are involved in more than one organization at any given time. These outside interests and commitments inevitably constrain the behavior of participants in any given organization and, in some instances, strongly influence it. To regard participants as completely constrained by the organization is to misperceive one of the fundamental characteristics of modern organizations: that they are systems built on the partial involvement of their members" (21–22).

7. Kesey, *Cuckoo's Nest*, 33.

8. Similarly, in "Race War Within," Medovoi has argued that texts as seemingly disparate as Heinlein's *Puppet Masters* and Ginsberg's "Howl" are "structured by an analogous defense of human freedom against the onslaught of totalitarianism" (174).

9. Ellison, *Invisible Man*, 505.

10. Foucault, *Discipline and Punish*, 238.

11. Erving Goffman, *Asylums: Essays on the Social Situation of Mental Patients and Other Inmates* (Chicago: Aidine, 1961), 43. Goffman's descriptions of the dynamics of these institutions steer clear, unlike those of Fromm or Lifton, from automaton imagery and are rich in their descriptions of social dynamics. For a recent introduction to Goffman and his methods of description, see David J. Alworth, "Melville in the Asylum," *ALH* 26, no. 2 (2014): 234–61. Alworth there notes that Goffman did not in fact coin the term *total institution*, though he did popularize it. I will add, too, that in a 1956 Macy Conference on

"Group Processes," Goffman presented an early version of the chapter "Characteristics of Total Institutions" alongside a presentation by Robert Jay Lifton on "Chinese Communist Thought Control." See Erving Goffman, "Interpersonal Persuasion," in *Group Processes: Transactions of the Third Conference, October 7, 8, 9, 10, 1956,* ed. Bertram Schaffner, 117–93 (New York: Josiah Macy Foundation, 1956).

12. Betty Friedan, *The Feminine Mystique,* Norton Critical ed., ed. Kirsten Fermaglich and Lisa M. Fine (New York: W. W. Norton, 2013).

13. Such a project would follow in Bruno Latour's direction from "panopticon to oligopticon" in *Reassembling the Social: An Introduction to Actor Network Theory* (New York: Oxford University Press, 2008), 175, and Lily Kay's view of institutional collectives as "a kind of agency that both enables and constrains," as she describes it in *Who Wrote the Book of Life: A History of the Genetic Code* (Palo Alto, Calif.: Stanford University Press, 2000), xviii.

14. Ellison, *Invisible Man,* 94.

15. Stanley Cavell, *The Claim of Reason: Wittgenstein, Skepticism, Morality, and Tragedy* (New York: Oxford University Press, 1979), 403–18. For Cavell, the ethical gesture of "acknowledgment" is prior to all forms of scientific and philosophical knowledge. A similar assumption grounds Ellison's attempt to undermine forms of white scientific knowledge about African Americans that explicitly withhold such acknowledgment.

16. Such critical accounts include Houston Baker's *Blues, Ideology, and Afro-American Literature* (Chicago: University of Chicago Press, 1984) and Henry Louis Gates Jr.'s *The Signifying Monkey* (New York: Oxford University Press, 1988), as well as contextualizations within the postwar culture of anticommunism (Barbara Foley), the organization man (Andrew Hoberek, discussed later), and postwar intellectuals (Jerry Gafio Watts and Kenneth Warren). My work also builds on, though ultimately disagrees with, other recent work that has paid attention to the novel's automatons. Sianne Ngai, *Ugly Feelings* (Cambridge, Mass.: Harvard University Press, 2005), suggests in a brief reading of the Sambo doll scene (discussed later) that the representation of automatons is related to a crisis of agency associated with African American "animatedness . . . a representation of the African American . . . as excessively 'lively' and a pliant body" (12). Bill Brown, whose work I also discuss later, has read some of the novel's scenes with automatons as part of a longer discussion of racist objects and memorabilia.

17. I understand Ellison's thinking about how African Americans do or do not "count" as part of a democratic community through Jacques Rancière's writings, discussed briefly in the introduction. He posits that reorganizing the count of citizens is the fundamental gesture of democratization: "political dispute is distinct from all conflicts of interest between constituted parties of the population, for it is a conflict over the very count of those parties. It is not

a discussion between partners but an interlocution that undermines the very situation of the discussion." Rancière, *Disagreement*, 100.

18. Ellison, *Invisible Man*, 235.

19. Ibid., 237.

20. Papers of Ralph Waldo Ellison, box 146, folder 13, Manuscript Division of the Library of Congress, Washington, D.C. This folder was retyped in 1949, suggesting that this phrase was cut relatively late in the writing process. Aside from being a somewhat feeble pun, this utterance may have been excised because it is spoken in this draft by an earlier incarnation of the character "Mary," who will be cast in a different role in the final version of the novel. In this version, she is a kindly janitor who aids the narrator in escaping the factory hospital.

21. By the time he would write *Invisible Man*, Ellison had already incorporated what is now a commonplace mode of satirical asylum imagery into his fiction, using the figures of the straitjacket-toting guards as lackeys of the racist power structure in his 1944 short story "Flying Home": these guards have the power to declare a well-to-do African American fighter pilot "insane" for reaching above his "natural" position. A similar satirical mode is taken up again in *Invisible Man*, when the dean of the Negro college, Bledsoe, has a war veteran shipped away to another mental institution after speaking too freely to the white philanthropist Norton (151–52). In this moment, Ellison combines the notion of the mental institution as a space of containment for radical or outrageous ideas with a critique of the black bourgeoisie, embodied in Bledsoe, as the social group guilty of such silencing. Nevertheless, the novel's factory hospital scene seems not to fit within even this transposed version of the familiar Foucauldian critique of mental institutions. Rather, despite the presence of electroconvulsive therapy as an instance of institutional violence, the "case" that has been developing for "three hundred years" demands a critical frame capable of understanding the reach of sociology's implications, comprising deeper questions about the act of scientific interpretation.

22. See Daniel Kim, *Writing Manhood in Black and Yellow: Ralph Ellison, Frank Chin, and the Literary Politics of Identity* (Stanford, Calif.: Stanford University Press, 2005), 47, and Roderick Ferguson, *Aberrations in Black: Toward a Queer of Color Critique* (Minneapolis: University of Minnesota Press, 2004), 71–76. Ferguson uncovers from Ellison's papers a discarded chapter of *Invisible Man* that utilizes Robert Park's infamous statement that "the Negro is the lady of the races" and analyzes the gender politics of the Trueblood scene in the light of John Dollard's *Caste and Class in a Southern Town* (Garden City, N.Y.: Doubleday, 1957), a sociological study in the Chicago tradition. Stephen Schryer, *Fantasies of the New Class: Ideologies of Professionalism in Post–World War II American Fiction* (New York: Columbia University Press, 2011), discusses Ellison's engagement with Myrdal more centrally but claims that, as a member of a new professional–managerial class, Ellison is closer to Myrdal

than he realizes. Kenneth Warren also discusses the Myrdal review as a key moment in Ellison's intellectual development in *So Black and Blue: Ralph Ellison and the Occasion for Criticism* (Chicago: University of Chicago Press, 2003), 32–33.

23. Singh, *Black Is a Country*, calls *An American Dilemma* "the landmark reference work for the long civil rights era" (142). He argues that *An American Dilemma* is particularly important for its articulation of the international significance of Jim Crow: "In the final analysis, what made *An American Dilemma* such an influential document was its presentation of 'the Negro problem' as the symbolic pivot on which future claims to [the] U.S. global mission rested" (148).

24. Ibid., 134. Although Myrdal himself had great hopes that his study would help to bring about racial justice in the United States, the fact that riot control forms part of the project's original impetus adds another layer of significance to the riot scene in Ellison's novel.

25. The question of Ellison's own position relative to African American sociology is a complex one, in part because he mostly avoids direct engagements with it, perhaps out of politeness or a fear of alienating himself from the community of prominent African American intellectuals. Arnold Rampersad, *Ralph Ellison: A Biography* (New York: Knopf, 2007), notes that Ellison sent courtesy copies of *Invisible Man* as a way to fuel "special rivalries," and the list of recipients included both Richard Wright and the prominent black sociologist Horace Cayton (259). On a conceptual level, Ellison might have agreed with Marlon Ross's assessments in *Manning the Race* (New York: New York University Press, 2004) that African American sociologists "used urban ethnography to construct their own masculinity as normal, their sexuality as self-disciplined, and their social status as professional—that of men deserving managerial responsibility for the black urban mass" (147). I see Ellison's humanism as drawing on a tradition of African American definitions of cultural analysis as distinct from scientific analysis, including Du Bois's own statement, preceding his musicological analyses in "The Sorrow Songs," in *The Souls of Black Folk* (New York: Oxford University Press, 2007), that "so woefully unorganized is sociological knowledge that the meaning of progress . . . and the limits of human perfectability, are veiled, unanswered sphinxes on the shores of science" (192). Likewise, Alain Locke opens his essay "The New Negro," in *The New Negro: An Interpretation* (New York: Boni, 1925), with the claim that the titular character of his essay has remained invisible to the "watch and guard of statistics" kept by the "Sociologist, the Philanthropist, [and] the Race Leader" (3). Take Du Bois's oft-quoted line from *Souls of Black Folk*: "while sociologists gleefully count his bastards and his prostitutes, the very soul of the toiling, sweating black man is darkened by the shadow of a vast despair" (42). It introduces a useful distinction between the object of the count (the bastard) and the "very soul" it neglects, a distinction that Ellison uses in his depiction of the Brotherhood. In this line of thinking, one might

argue that Ellison's disavowal of sociological knowledge (which he would likely have associated with Wright) can be seen as an enabling condition for his faith in the promise of cultural analysis and action.

26. See Herman, *Romance of American Psychology*, 182. Herman claims that Adorno's text is the first to focus on the question of prejudice as "determined by deep psychic structures," a model that Myrdal embraces. As for the efficacy of this method, I would argue that Ellison, like Horton, rejects outright Myrdal's belief that exposing the contradiction of the Jim Crow practice with the ideal of the "American creed" will be, in itself, all that is needed to solve the "Negro problem."

27. Ellison, *"An American Dilemma*: A Review," in *The Collected Essays of Ralph Ellison*, ed. John Callahan (New York: Modern Library, 1995), 339; Gunnar Myrdal, *An American Dilemma*. It appears that Ellison misquoted this passage from his notes, though he does not misrepresent the gist of Myrdal's argument. Ellison's quotation appears to be a paraphrase of one or more of the following passages: (1) "History is never irredeemable, and there is still time to come to good terms with colored peoples. Their race pride and race prejudice is still mostly a defensive mental device, a secondary reaction built up to meet the humiliations of white supremacy" (1018); (2) "The voluntary withdrawal and the self-imposed segregation were shown to be a secondary reaction to a primary white pressure" (669n1); or (3) "Negro thinking is almost completely determined by white opinions—negatively and positively. It develops as an answer to the popular theories prevalent among whites by which they rationalize their upholding of caste. In this sense it is a derivative, or secondary, thinking. The Negroes do not formulate the issues to be debated; these are set for them by the dominant group" (784). It would seem that the variety of similar statements throughout Myrdal's work serves to underscore the centrality of this idea of a "secondary reaction" to Myrdal's methodology.

28. Myrdal, *An American Dilemma*, 781–86. Singh's account of Myrdal's representations of African American politics in *Black Is a Country* agrees with Ellison's impression of the work: "In the end, Myrdal denied the autonomous capacity of black people as individuals and as a collectivity (even as humans) to formulate a coherent, public standpoint on the social and political realities of American life" (147).

29. Ellison, *"An American Dilemma*: A Review," 339.

30. Carol Horton, *Race and the Making of American Liberalism* (New York: Oxford University Press, 2005), 123.

31. Ibid.

32. Ellison, *"An American Dilemma*: A Review," 337–38.

33. Rampersad, *Ralph Ellison: A Biography*, 181. See also Jerry Gafio Watts, *Heroism and the Black Intellectual: Ralph Ellison, Politics, and Afro-American Intellectual Life* (Chapel Hill: University of North Carolina Press, 1994). Reading Ellison's essays independently of his fiction, Watts claims that Ellison's

response to sociology amounts only to "one-dimensional proclamations of black human agency against one-dimensional social-scientific denials of such agency" (57). I claim that, by reading his work on sociology in conjunction with his fiction, it is clear that Ellison puts a great deal of thought into how that denial of agency functions and into ways of recuperating that agency, which I discuss in the conclusion.

34. Ellison, *Invisible Man*, 237.

35. See Martin Heidegger, *Being and Time,* trans. John Macquarrie and Edward Robinson (New York: Harper, 1962), esp. para. 14, and also Hubert Dreyfus, *Being-in-the-World: A Commentary on Heidegger's Being and Time, Division I* (Cambridge, Mass.: MIT Press, 1991), 29–33.

36. Martin Heidegger, "The Age of the World-Picture," in *The Question Concerning Technology, and Other Essays* (New York: Harper and Row, 1977), 21.

37. See also Fred Moten, "The Case of Blackness," *Criticism* 50, no. 2 (2008): 177–218. Moten makes the suggestive point that the work of Frantz Fanon, in *Black Skin, White Masks,* trans. Charles Lam Markmann (New York: Grove Press, 1967), "places the Heideggerian distinction between being (thing) and Dasein—the being to whom understandings of being are given; the not, but nothing other than, human being—in a kind of jeopardy that was already implicit" (186). I would suggest that Ellison's automatons place a similar kind of pressure on this distinction, but that, for both Ellison and Fanon, the fact of being seen only as object calls for a kind of redress, in the form of an attempt to convey one's psychological complexity to the scientist or, as in Fanon's essay, to the terrified young white girl on the street.

38. Seltzer, *Bodies and Machines*, 100.

39. Ellison, *Invisible Man,* 237.

40. In a recent essay, Brown, "Reification, Reanimation, and the American Uncanny," has pointed critical attention toward automaton figurines of African Americans, in *Invisible Man* and Spike Lee's *Bamboozled,* where he concludes that African American "golliwog" memorabilia often evoke the United States's slaveholding past as a point of unsettling confusion between people and objects. Brown cites in particular "the apprehension that within things we will discover the human precisely because our history is one in which humans were reduced to things (however incomplete that reduction)" (207). In considering Ellison's novel as an "elaborate organization of both plot and character as a series of object-relations" (202), Brown enables a particularly productive approach to the novel. I would contend, however, that it is not only America's slaveholding past but also a new set of phenomena that weigh on American consciousness, namely, the construction by scientific, bureaucratic, and state institutions of the African American as a product of the Negro problem.

41. Ellison, *Invisible Man,* 319.

42. This moment in the novel also resonates with an earlier instance in the "battle royal" scene, in which the narrator is made to grab for coins on an electric carpet: in addition to the repeated image of African American bodies receiving electric shocks, the narrator sees "attendants in white jackets" put the electrified rug in place. The white jackets, suggestive of the lab coats worn by medical doctors and many research scientists, echo with the doctors who go on to discuss the narrator's "case" in the factory hospital, and the earlier scene with the carpet is also explicitly framed as a piece of entertainment for white audience. Ibid., 26. Through this parallel, the novel not only links these scientists to the broad, degrading physical humor that defined racist depictions of African Americans of the pre–civil rights United States but also calls on the reader to see the scientist's white jacket as a symbol of a desire to repress racial threats in a tidy fashion.

43. Ellison, *Invisible Man,* 236.

44. Here I would argue against critics like Barbara Foley, "The Rhetoric of Anticommunism in *Invisible Man,*" *College English* 59 (1997): 530–47, who accuses Ellison of inhabiting in a "rhetoric of anti-communism" to curry favor with white audiences (530). Ellison's is an anticommunism that criticizes the particular party approach to the social problem at hand, and with sufficient substance and continuity to suggest that the depiction of the Brotherhood was not a callous or superficial marketing calculation on Ellison's part. This argument is reprised in Foley, *Wrestling with the Left: The Making of Ralph Ellison's* Invisible Man (Durham, N.C.: Duke University Press, 2010), 283–302.

45. Andrew Ross, *No Respect: Intellectuals and Popular Culture* (New York: Routledge, 1989), 22. For an example of Marxian thinking after 1943, see Ellison's 1944 *"An American Dilemma:* A Review," where he claims that the "Negro problem" is "where Marx cries out for Freud and Freud for Marx," in addition to criticizing Myrdal for ignoring "the class struggle" (335, 339).

46. Most notably Arendt, in the 1958 second edition of *Origins of Totalitarianism,* which states the equivalence between Nazism and communism as their scientific motivation in forcing a "law of Nature" (Darwinism) or a "law of History" (Marxism) into being, which "claim[s] to transform the human species into an active carrier of a law to which human beings otherwise would only passively and reluctantly be subjected" (462).

47. Aldous Huxley, *Brave New World* (1932; repr., New York: Harper and Bros., 1946).

48. Adorno and Horkheimer, *Dialectic of Enlightenment,* x.

49. Ellison, *Invisible Man,* 306.

50. Ibid., 503.

51. Ibid., 382.

52. Ibid., 305.

53. This gives one plausible explanation for Barbara Foley's finding that the narrator's salary was not accurate for the Communist Party of the time:

earning more money makes the narrator's joining the Brotherhood primarily a means of economic success rather than a matter of being persuaded to join out of belief. In an early outline of *Invisible Man*, Ellison writes that he joins the Brotherhood "not because he believes, but because they ask him during his moment of deepest despair," though eventually he becomes "convinced that [in the Brotherhood] he has found a real democracy," Ellison, *Collected Essays*, 347. Another explanation for this way of including the Brotherhood material can be found in Ellison's files, where he had saved a clipping of a communist exposé from *Life* magazine titled "Portrait of an American Communist." It cites this man's salary as eighty-five dollars a week for fourteen hours of work per week, more than the narrator's salary of sixty dollars. It also includes a somewhat titillating account of interracial relationships in the party. Such an article might have shown Ellison that the communist exposé was a viable and sale-able genre of storytelling. Ralph Waldo Ellison Papers, box 200, folder 6; John McPartland, "Portrait of an American Communist," *Life*, January 5, 1948, 23–26.

54. Ralph Ellison, "Working Notes for Invisible Man," in *Collected Essays*, 346.

55. Ellison, *Invisible Man*, 312.

56. Ibid., 473.

57. Ibid., 507.

58. Ibid. The final scene before the epilogue, in fact, continues this association between the sociologists' scientism, the Brotherhood, and the narrator's automatism. The fact that it is a dream sequence, following the narrator's escape from the riot, seems to allow for the inclusion of an even stranger portrayal of the relationship between man, scientist, and automatism than has been allowed elsewhere in the novel. In a clear parallel with the scene in the factory hospital, the narrator finds himself prostrate, surrounded by men standing over him—this time, the main characters from his past, including Jack, Norton, Tobitt, Ras, and Bledsoe, rather than just the psychiatrists. In the dream, the men castrate him, fulfilling the suggested prescription of one of the doctors in the factory hospital scene midway through the novel. Jack asks the narrator how it feels "to be free of one's illusions," and the narrator sees his "blood-red parts" hanging in the arch of a high, metal bridge (569). As he points to the bloody spectacle, he tells them, "There hang not only my generations wasting upon the water [but] your universe, and that drip-drop upon the water is all the history you've made. . . . Now laugh, you scientists. Let's hear you laugh!" (570). His insistence that these men are "scientists" cements the connection, in the novel's fabric of repeated motifs, between the science of the psychiatrists and that of the Brotherhood. In this image, the bridge from which the narrator's genitals hang begins to walk, "striding like a robot, an iron man, whose iron legs clanged doomfully as it moved" (570). As a dream of castration anxiety, it may suggest that the narrator's fear is particularly well

founded, because the whites in the novel ascribe an unrealistic power or life force to the African American phallus. The suggestive image of the giant robot might also convey a mood of totalitarian horror, wherein the version of the narrator that the scientists create is also an ominous, inhuman, and destructive monstrosity.

59. Ralph Waldo Ellison Papers, box 143, folder 2, 667.

60. This portion of the conversation between Jack and the narrator may ultimately have been deleted for its similarity to what has become one of the most widely cited passages in *Black Boy,* Richard Wright's statement that African Americans "lived somehow in [Western civilization] but not of it." In Wright's powerful turn of phrase, the semantic difference between being "in" and "of" Western civilization signals a mode of alienation, or even of double consciousness, the modifier "somehow" carrying the weight of the senselessness of history.

61. Ellison, *Invisible Man,* 377.

62. See Brown, "Reification, Reanimation, and the American Uncanny," 207, and Ngai, *Ugly Feelings,* 12.

63. Ellison, *Invisible Man,* 372.

64. Ibid., 434.

65. Watts's excellent study of Ellison's essays, *Heroism and the Black Intellectual,* proposes a perspective on the Communist Party that is useful for interpreting this moment in biographical terms as well: Watts claims that, for both Ellison and Wright as authors, the Communist Party had been a "social marginality facilitator" (16), that is to say, that early on, both Ellison and Wright saw the party as one of the only white audiences willing to read their work (and, in Wright's case, they served as a source of education).

66. Ellison, *Invisible Man,* 431–32.

67. Ngai, *Ugly Feelings,* 113.

68. Ellison, *Invisible Man,* 432.

69. Ibid., emphasis original.

70. Ibid., 431.

71. Ibid., 433.

72. Ibid., 467.

73. Ibid.

74. Ibid., 464, 469.

75. Ibid., 470.

76. Ibid., 548.

77. Ibid., 452.

78. Ibid., 441.

79. Ibid., 380. The Hot Foot Squad, in its creative and corporeal aspects, is squarely in line with Andrew Ross's useful distinction in *No Respect* between the roles of creativity and the body in the Old Left as opposed to the New Left, which is marked by a new "attention to personal, liberatory values [as] a major

element in redefining responsibility in terms either addressed to the body directly, or else enlisted the mind and psyche as media of self-transformation, rather than as tools to be harnessed to objective political causes" (220).

80. Ellison, *Invisible Man*, 81.

81. Ralph Ellison, "Brave Words for a Startling Occasion," in *Collected Essays*, 151.

82. Ishmael Reed, *Mumbo Jumbo* (New York: Scribner, 1972), 17, and Janelle Monáe, *Metropolis: The Chase Suite* (New York: Atlantic Records, 2008); Monáe, *The ArchAndroid* (New York: Atlantic Records, 2010), and Monáe, *Electric Lady* (New York: Atlantic Records, 2013). On Reed's use of the robot figure and its connections with cybernetics, see Michael Chaney, "Slave Cyborgs and the Black Infovirus: Ishmael Reed's Cybernetic Aesthetics," *MFS: Modern Fiction Studies* 49, no. 2 (2003): 261–83.

83. Charles A. Reich, *The Greening of America* (New York: Bantam, 1970), 129–30.

84. For a history of the subliminal message, see Charles Acland, *Swift Viewing: The Popular Life of Subliminal Influence* (Durham, N.C.: Duke University Press, 2012).

85. Friedan, *Feminine Mystique*, 174.

86. Ibid., 178.

87. Ibid.

88. Ibid., 256.

89. Arendt, *Origins of Totalitarianism*, xxxii.

90. Louis Althusser, a decade later, would include the family as a primary institution in his list of ideological state apparatuses in "Ideology and Ideological State Apparatuses," trans. Ben Brewster, in *Lenin and Philosophy and Other Essays*, 121–76 (London: New Left Books, 1971). Fredric Jameson's *The Political Unconscious: Narrative as a Socially Symbolic Act* (Ithaca, N.Y.: Cornell University Press, 1981) would be a primary vehicle for Althusser's thought to enter the U.S. academy in the 1980s and 1990s, and alongside Foucault's *Discipline and Punish*, it would enable literary and cultural studies scholarship to expand the form of the institution further into critical accounts of the state, patriarchy, racism, and more.

91. Kirsten Lise Fermaglich has argued that this comparison also provided Friedan with a means of rejecting Cold War consensus politics in two ways: first, she "dissented from the Cold War consensus that glorified women's domesticity as a means of battling the Soviet Union," and second, as a suggestion of "similarities between Nazi Germany and the United States," Friedan also questions the concept of the totalitarianism that links Nazi Germany to the USSR. Fermaglich, "'The Comfortable Concentration Camp': The Significance of Nazi Imagery in Betty Friedan's *The Feminine Mystique* (1963)," *American Jewish History* 91, no. 2 (2003): 223.

92. Friedan, *Feminine Mystique*, 254. The historian Susan Lisa Carruthers, in *Cold War Captives*, 207–9, also includes Friedan's metaphor as part of a larger framework of "captivity" as a metaphor during the Cold War.

93. In *Feminine Mystique*, Friedan cites Riesman's observation of "a basic change in the American character" after World War II, and particularly a sort of "passivity" in American young people that had been likened to the "deterioration of the human character" experienced by the "American GI's who were prisoners of war in Korea in the 1950s" (236, 237). She turns this rhetoric around quite brilliantly, writing that "the apathetic, dependent, infantile, purposeless being, who seems so shockingly nonhuman when remarked as the emerging character of the new American man, is strangely reminiscent of the familiar 'feminine' personality as defined by the [feminine] mystique" (238).

94. See Carruthers, *Cold War Captives*, for a lengthier discussion of the metaphor of captivity in Friedan's work.

95. Friedan, *Feminine Mystique*, 254.

96. Ibid., 257.

97. Ibid.

98. Ibid., 305, emphasis original.

99. Ibid. One housewife Friedan interviews earlier in the text puts that plan in terms of having the right machines: "It's nice to be modern—it's like running a factory in which you have all the latest machinery" (211).

100. Ira Levin, *The Stepford Wives* (1972; repr., New York: William Morrow, 2002). Levin's first two novels were both made into films: *A Kiss before Dying* (1953) in 1956 (and 1991) and *Rosemary's Baby* (1967) in 1968. His contribution to the automaton horror genre was almost certainly written with film adaptation in mind.

101. Levin, *Stepford Wives*, 37.

102. Cf. Erich Fromm, citing interviews about the Park Forest suburb, in *The Sane Society* (New York: Routledge, 1956): "What is important in this picture is not only the fact of alienated friendships, and automaton conformity, but the reaction of people to this fact. Consciously it seems people fully accept the new form of adjustment. Once people hated to concede that their behavior was determined by anything except their own free will. Not so with the new suburbanites; they are fully aware of the all-pervading power of the environment over them" (161).

103. Anna Krugovy Silver, in "The Cyborg Mystique: *The Stepford Wives* and Second Wave Feminism," *Arizona Quarterly* 58, no. 1 (2002): 109–26, collects the 1970s responses to the film from outlets including *Time, Newsweek,* and *The New Yorker,* with the overview that "critics of *The Stepford Wives* either derided it, on the one hand, as anti-male or, on the other hand, as a misrepresentation of feminist goals and cultural critiques" (110). Silver herself posits that "the film . . . suggests not the failure and perversion of feminist rhetoric, as Friedan implies [in her disparaging 'rip-off' remark], but its success and

popular appeal" (110). Sherryl Vint has observed, in "The New Backlash: Popular Culture's 'Marriage' with Feminism, or, Love Is All You Need," *Journal of Popular Film and Television* 34, no. 4 (2007): 160–69, that "the Men's Association members are nerdy losers, but the film nonetheless drifts dangerously close to endorsing their adolescent fantasies, as it is Joanna, rather than the men, who must suffer through the wives' insipidness" (164).

104. On branding, see Naomi Klein, *No Logo: No Space, No Choice, No Jobs* (New York: Picador, 1999).

105. Thomas Frank, *The Conquest of Cool: Business Culture, Counterculture, and the Rise of Hip Consumerism* (Chicago: University of Chicago Press, 1998), 4. Frank's otherwise definitive account does not address Apple's "1984" ad.

106. The Macintosh was the first GUI computer, but not the first personal computer, which began to be marketed and sold in the late 1970s. On the transition from the mainframe era to the PC era, see Alan Liu, *The Laws of Cool: Knowledge Work and the Culture of Information* (Chicago: University of Chicago Press, 2004), 142.

107. In ibid., Liu also explores this anti-institutional paradox with regard to corporate knowledge work more generally. In his articulation of the Silicon Valley mantra "We work here, but we're cool," a 1960s-style distance from "the man" and the soullessness of corporations became a mode of self-understanding even with those corporations themselves (76).

3. Human Programming

1. The literary works of figures like Kurt Vonnegut, William S. Burroughs, and Neal Stephenson have occupied spaces in critical literature and fan culture between the poles of science fiction, avant-garde literature, and literary fiction. Vonnegut's *Player Piano* is a strong example of this betweenness: originally released by Scribner in 1952, it was rebranded as science fiction in 1954, presumably to increase its sales, and released as *Utopia 14* (New York: Bantam Books, 1954).

2. McGurl, *Program Era*, 68.

3. "Instead of accepting cyberpunk's own promotional mythology of its difference from earlier science fiction . . . it might be better historicized as an intervention in a set of other posthuman discourses and speculative cultures." Thomas Foster, *The Souls of Cyberfolk: Posthumanism as Vernacular Theory* (Minneapolis: University of Minnesota Press, 2005), xii. See also Veronica Hollinger, "Cybernetic Deconstructions: Cyberpunk and Postmodernism," *Mosaic* 23 (Spring 1990): 29–44.

4. Norbert Wiener recounts this interaction in *Cybernetics; or, Control and Communication in the Animal and the Machine* (Cambridge, Mass.: MIT Press, 1948), 28.

5. Daniel Bell's book *Coming of the Post-industrial Society: A Venture in Social Forecasting* (New York: Basic Books, 1976) is the canonical prediction of

this form of society, wherein, "if industrial society is based on machine technology, post-industrial society is shaped by intellectual technology . . . information and knowledge" (xv). Thomas Piketty estimated that in 2012, approximately 80 percent of the U.S. workforce would be classified as "service," as opposed to "manufacturing" (18 percent) or "agriculture" (2 percent), in *Capital in the Twenty-First Century* (Cambridge, Mass.: Harvard University Press, 2013), 90.

6. Kurt Vonnegut, *Player Piano* (New York: Scribner, 1952).

7. Ibid., 11.

8. Liu, *Laws of Cool*, 117.

9. Wiener, *Cybernetics*, 19–21.

10. Norbert Wiener, "Men, Machines, and the World About," in *The New Media Reader*, ed. Noah Wardrip-Fruin and Nick Montfort (Cambridge, Mass.: MIT Press, 2003), 68.

11. Thomas Huxley's essay "On the Hypothesis That Animals Are Automata," *Fortnightly Review* 16, no. 95 (1874): 555–80, has a plausible claim to priority here in that the conscious automatism he ascribes to animals is also applicable to humans. Descartes and Aristotle also compared animals to machines in their writings. Wiener's fullest explanation of the human–machine comparison came in *The Human Use of Human Beings: Cybernetics and Society* (1950; repr., Boston: Da Capo Press, 1988): "Thus the nervous system and the automatic machine are fundamentally alike in that they are devices which make decisions on the basis of decisions they have made in the past. The simplest mechanical devices will make decisions between two alternatives, such as the closing or opening of a switch. In the nervous system, the individual nerve fiber also decides between carrying an impulse or not. In both the machine and the nerve, there is a specific apparatus for making future decisions depend on past decisions, and in the nervous system a large part of this task is done at those extremely complicated points called 'synapses' where a number of incoming nerve fibers connect with a single outgoing nerve fiber" (33–34).

12. See Niklas Luhmann, *Social Systems*, trans. John Bednarz Jr., with Dirk Baecker (Stanford, Calif.: Stanford University Press, 1995), and F. G. Varela, H. Maturana, and R. Uribe, "Autopoiesis: The Organization of Living Systems, Its Characterization and a Model," *Biosystems* 5 (1974): 187–96.

13. Wiener, *Cybernetics*, 148. It's notable that Wiener, even in these circumstances, where he compares humans to machines, keeps the humane aspects of humanness ever present in his work. He describes the "more violent" intervention of the prefrontal lobotomy and its potential for "permanent damage," and he remarks sardonically on its "current vogue" as a facilitator of custodial care. "Let me remark in passing that killing them makes their custodial care still easier," Wiener writes in a characteristically acerbic note (148).

14. Wiener, *Human Use of Human Beings*, 51.

15. John von Neumann, *The Computer and the Brain* (New Haven, Conn.: Yale University Press, 2012), 8, 51.

16. David Golumbia, *The Cultural Logic of Computation* (Cambridge, Mass.: Harvard University Press, 2009), 31–129.

17. Wendy Hui Kyong Chun, *Programmed Visions: Software and Memory* (Cambridge, Mass.: MIT Press, 2011), 9.

18. See Kay, *Who Wrote the Book of Life,* and Judith Roof, *The Poetics of DNA* (Minneapolis: University of Minnesota Press, 2007). For discussions of genetics in fiction, see also Priscilla Wald, "Future Perfect," *New Literary History* 31, no. 4 (2000): 681–708; Gerry Canavan, "Life without Hope? Huntington's Disease and Genetic Futurity," in *Disability in Science Fiction: Representations of Technology as Cure,* ed. Kathryn Allan, 169–88 (New York: Palgrave Macmillan, 2013); and Jay Clayton, "The Ridicule of Time: Science Fiction, Bioethics, and the Posthuman," *ALH* 25, no. 2 (2013): 317–43. I pursue this distinction between programmable minds and bodies within a different line of argument in Scott Selisker, "GMOs and the Aesthetics of Scale in Paolo Bacigalupi's *The Windup Girl,*" *Science Fiction Studies* 42, no. 3 (2015): 500–518.

19. Andrew Niccol, *Gattaca* (Culver City, Calif.: Columbia TriStar Home Video, 1998). On the "gay gene," see Roof, *Poetics of DNA,* 136–37.

20. Jay Clayton has found in the "waves" of fiction about genetics, both in the eugenic fictions of the 1950s and in transgenic fictions since the 1970s, a "typical plot form" that stages "the persecution of the emerging minority species by a terrified majority, the soon-to-be extinct *Homo sapiens,*" in "Ridicule of Time," 321.

21. William S. Burroughs, *Nova Express* (New York: Grove Press, 1994).

22. Ibid., 56, emphasis original.

23. Ibid.

24. Ibid., 72.

25. Wald, *Contagious,* 184–89; Seed, *Brainwashing,* 134–56. Wald connects Burroughs with virology, whereas Seed connects him to cybernetics and mind control discourses.

26. Shulamith Firestone, *The Dialectic of Sex: The Case for Feminist Revolution* (1970; repr., New York: Farrar, Strauss, and Giroux, 2003).

27. Ibid., 200.

28. Nancy Hartsock, "The Feminist Standpoint: Developing the Ground for a Specifically Feminist Historical Materialism," in *Discovering Reality: Feminist Perspectives on Epistemology, Metaphysics, Methodology, and Philosophy of Science,* ed. Sandra Harding and Merrill B. Hintikka (Dordrecht, Netherlands: D. Reidel, 1983), 293.

29. Valerie Solanas, *SCUM Manifesto,* ed. Avital Ronell (New York: Verso, 2004), 35.

30. Firestone, *Dialectic of Sex,* 234, 11.

31. Ibid., 201.

32. Arlie Russell Hochschild, *The Managed Heart: Commercialization of Human Feeling* (Berkeley: University of California Press, 1983).

33. Ibid., 142.

34. Liu, *Laws of Cool*, 436.

35. Hochschild, *Managed Heart*, 38.

36. Firestone, *Dialectic of Sex*, 90, emphasis original.

37. Philip K. Dick, *Do Androids Dream of Electric Sheep?* (1968; repr., New York: Ballantine Books, 2008).

38. Hayles, *How We Became Posthuman*, 2.

39. For a recent review and discussion of *Do Androids Dream* alongside Masahiro Mori and the Turing test, see Jennifer S. Rhee, "Beyond the Uncanny Valley: Masahiro Mori and Philip K. Dick's *Do Androids Dream of Electric Sheep?*," *Configurations* 21, no. 3 (2013): 301–29.

40. Dick, *Do Androids Dream*, 38.

41. Ibid., 103.

42. Here I am borrowing from and adapting distinctions set forth in the work of Gilbert Simondon and Brian Massumi; see Brian Massumi, *Parables for the Virtual: Movement, Affect, Sensation* (Durham, N.C.: Duke University Press, 2002), 27, 294. Hochschild also describes the history of theories of emotion, in which organismic models (Darwin, William James, Freud) were taken over by interactionist ones (John Dewey, C. Wright Mills, Erving Goffman) in the mid-twentieth century; see Hochschild, *Managed Heart*, 216–17.

43. Moreover, as Hayles notes in *How We Became Posthuman*, 177, emotional presence and the notion of empathy also become suspended between humans and electronic interfaces in *Do Androids Dream*.

44. Isaac Asimov, "Evidence," in *I, Robot* (1946; repr., New York: Spectra, 2004), 170–197; Asimov, "The Bicentennial Man," in *The Bicentennial Man and Other Stories* (New York: Doubleday, 1976).

45. Spike Jonze, *Her* (Burbank, Calif.: Warner Home Video, 2014); Caradog James, *The Machine* (North Hollywood, Calif.: Xlrator Media, 2014); Gabe Ibáñez, *Autómata* (Los Angeles, Calif.: Millenium Entertainment, 2014); Alex Garland, *Ex Machina* (Santa Monica, Calif.: Lionsgate, 2015).

46. Stanley Cavell, *The Claim of Reason: Wittgenstein, Skepticism, Morality, and Tragedy* (New York: Oxford University Press, 1979), 405–15.

47. Sherryl Vint, *Bodies of Tomorrow: Technology, Subjectivity, Science Fiction* (Toronto: University of Toronto Press, 2007), 190.

48. Dick, *Do Androids Dream*, 158.

49. Philip K. Dick, "The Android and the Human," in *The Shifting Realities of Philip K. Dick*, ed. Lawrence Sutin (New York: Pantheon Books, 1995), 194–95.

50. Philip K. Dick, "Second Variety," in *Space Science Fiction* 1, no. 6 (1953): 102–45; Dick, *The Simulacra* (New York: Ace Books, 1964); Dick, *The Three Stigmata of Palmer Eldritch* (Garden City, N.Y.: Doubleday, 1965).

51. Greif, *Age of the Crisis of Man*, approaches my own claim about the relationship between democratic selves and totalitarian others when he describes

discourse about the "dignity of man," which "could be made a name for whatever was good about American democracy and bad about the USSR, since one
system (democracy) knew what people were 'really like' and the other (authoritarian socialism) betrayed human dignity" (12).

52. Greta Aiyu Niu, "Techno-Orientalism, Nanotechnology, Posthumans,
and Post-Posthumans in Neal Stephenson's and Linda Nagata's Science Fiction," *MELUS* 33, no. 4 (2008): 73–96. See also David Roh, Greta Niu, and
Betsy Huang, eds., *Techno-Orientalism in Science Fiction Film, Media, and Literature* (New Brunswick, N.J.: Rutgers University Press, 2015).

53. David S. Roh, Betsy Huang, and Greta A. Niu, "Technologizing Orien-
talism: An Introduction," in *Techno-Orientalism: Imagining Asia in Speculative
Fiction, History, and Media,* ed. David S. Roh, Betsy Huang, and Greta A. Niu
(New Brunswick, N.J.: Rutgers University Press, 2015), 2.

54. Abigail De Kosnick also notes that George Lucas has mentioned the
Vietnam War as a historical framework for the Rebellion, and that Lucas borrowed costuming ideas and even plot structuring elements from samurai films,
in *"The Mask of Fu Manchu, Son of Sinbad,* and *Star Wars IV: A New Hope*:
Techno-Orientalist Cinema as Mnemotechnics of Twentieth-Century U.S.–
Asian Conflicts," in Roh et al., *Techno-Orientalism,* 98. In such a human–
inhuman dynamic between the Asian-mysticism-infused Rebellion and the
techno-totalitarian Empire (in the film's Riefenstahl-esque staging for its
stormtroopers and Imperial officers), it seems that, even though fraught with
stereotypes, the Asian-mysticism position is favored.

55. Roh et al., "Introduction," 5.

56. Lisa Lowe, *Immigrant Acts: On Asian American Cultural Politics* (Durham, N.C.: Duke University Press, 1996), 84–86.

57. Tsarina T. Prater and Catherine Fung, "'How Does It Not Know What It
Is?': The Techno-Orientalized Body in Ridley Scott's *Blade Runner* and Larissa
Lai's *Automaton Biographies,*" in Roh et al., *Techno-Orientalism,* 193–208.

58. See Larissa Lai, "Rachel," in *So Long Been Dreaming: Postcolonial Science
Fiction and Fantasy,* ed. Nalo Hopkinson and Uppinder Mehan, 53–60 (Vancouver, B.C.: Arsenal Pulp Press, 2004); Lai, *Automaton Biographies* (Vancouver,
B.C.: Arsenal Pulp Press, 2009); Karen Tei Yamashita, *Anime Wong* (Minneapolis, Minn.: Coffee House Press, 2014); and Greg Pak, "Machine Love," a
short in the film *Robot Stories* (New York: Pak Film, 2003).

59. Wolfe, *What Is Posthumanism?,* xiii.

60. Ibid., xv.

61. Mark McGurl, in an article on H. P. Lovecraft, "The Posthuman Comedy," *Critical Inquiry* 38, no. 3 (2012): 533–53, uses the term *posthuman* in the
sense of imagining frames of "deep time" that stretch either before the beginning of, or after the end of, humanity. Such an approach aligns especially well
with critical focuses on the "nonhuman" and the "anthropocene," whose critical gesture is to decenter humanity in its widest sense by imagining agency

beyond humans. For a fully gender-centered usage of *posthuman,* see Judith M. Halberstam and Ira Livingston, eds., *Posthuman Bodies* (Bloomington: Indiana University Press, 1995).

62. Haraway, "A Cyborg Manifesto," 149–82. Haraway identifies what I have been referring to as the "human automaton" when she notes that "from the seventeenth century till now, machines could be animated . . . or organisms could be mechanized—reduced to body understood as resource of mind" (178).

63. Ibid., 151.

64. Anne McCaffrey, *The Ship Who Sang* (New York: Ballantine, 1969).

65. Vonda McEntyre, *Superluminal* (Boston: Houghton Mifflin, 1983).

66. Haraway, "A Cyborg Manifesto," 178.

67. *The Bionic Woman,* perf. Lindsay Wagner, Richard Anderson, Harve Bennett Productions (Hollywood: Universal Studios Home Entertainment, 2010).

68. Hayles, *How We Became Posthuman,* 6.

69. Ibid., 279.

70. Ibid., 277.

71. Ibid., 278.

72. See Fredric Jameson, "After Armageddon: Character Systems in *Dr. Bloodmoney,*" *Science Fiction Studies* 2, no. 1 (1975): 31–42. My own diagram is not a proper Greimas square in that its four initial points are not derived from a single concept and its three varieties opposite; the shape of the schema is nevertheless useful for making a two-dimensional distinction between bodies and minds, organic and mechanical.

73. Hayles points out that the name resembles "such initialized luminaries as L. Ron Hubbard, L. B. J., and H. Ross Perot" in *How We Became Posthuman,* 273.

74. Stephenson, *Snow Crash,* 97.

75. Richard Dawkins, *The Selfish Gene* (New York: Oxford University Press, 1976), 57.

76. Ibid., 190.

77. For more elaboration on the distinction between mediators and intermediaries, see Latour, *Reassembling the Social,* 37–40. The initial articulation in cybernetics of information as a generalizable principle is Claude Shannon, "A Mathematical Theory of Communication," *The Bell System Technical Journal* 27 (July–October 1948): 379–423, 623–56.

78. Stephenson, *Snow Crash,* 200.

79. Ibid., 386.

80. Richard Rorty, *Achieving Our Country: Leftist Thought in Twentieth-Century America* (Cambridge, Mass.: Harvard University Press, 1998), 5.

81. Walter Benn Michaels, *The Shape of the Signifier: 1967 to the End of History* (Princeton, N.J.: Princeton University Press, 2004), 69. Michaels takes up *Snow Crash* in the course of a long disagreement with Rorty, and both authors

also discuss Leslie Marmon Silko's *Almanac of the Dead,* a novel in which Michaels's (to me, more convincing) argument about the primacy of blood over belief helps to explain his unusual approach to *Snow Crash.*

82. Foster, *Souls of Cyberfolk,* argues in a different register along a similar line: "*Snow Crash* demonstrates how Hayles's condition of virtuality may only invert the categories of nature and culture; as material bodies become more malleable, new forms of cultural rigidity are produced" (231). Within my reading of the novel, that line of thinking applies to the malleability of multiculturalism and the imagined rigidity of fundamentalism as much as to the status of virtual bodies in cyberspace.

4. Cult Programming

1. *Dollhouse,* created by Joss Whedon, perf. Eliza Dushku (Hollywood: Twentieth Century Fox Home Entertainment, 2009).

2. See Kenneth G. C. Newport, *The Branch Davidians of Waco: The History and Beliefs of an Apocalyptic Sect* (New York: Oxford University Press, 2006).

3. *Dollhouse's* episodes are often structured around real-life situations of programming or performance, and its "Epitaph" episodes stage a full-fledged apocalyptic scenario in which a human programming technology gets out of hand. On *Dollhouse's* relationship to zombie narratives, see Gerry Canavan, "Fighting a War You've Already Lost: Zombies and Zombis in *Firefly/Serenity* and *Dollhouse,*" *Science Fiction Film and Television* 4, no. 2 (2011): 173–204.

4. Mark Twain, *Christian Science* (New York: Harper, 1907).

5. Anthony A. Hoekema, *The Four Major Cults: Christian Science, Jehovah's Witnesses, Mormonism, Seventh-Day Adventism* (Grand Rapids, Mich.: Eerdmans, 1963).

6. William J. Peterson, *Those Curious New Cults* (New Canaan, Conn.: Keats, 1973).

7. For an overview of this literature's features, see Louise Stewart, "'Be Afraid or Be Fried': Cults and Young Adult Apocalyptic Narratives," *Children's Literature Association Quarterly* 36, no. 3 (2011): 318–35. According to Stewart, these narratives about young people often "rely on physical seductions combined with the seduction of spiritual promise" (322), and they sometimes also imagine the cult as an intensification of power dynamics within families (333).

8. See Zablocki, "Towards a Demystified and Disinterested Scientific Theory of Brainwashing," in *Misunderstanding Cults: Searching for Objectivity in a Controversial Field,* ed. Benjamin David Zablocki and Thomas Robbins, 159–214 (Toronto, Ont.: University of Toronto Press, 2001), for a thorough argument for the use of the term; for an equally thorough argument against, see Dick Anthony, "Tactical Ambiguity and Brainwashing Formulations: Science or Pseudo Science," in ibid., 215–317. David Bromley compares the rhetoric of "brainwashing" and "conversion" in "A Tale of Two Theories: Brainwashing and Conversion as Competing Political Narratives," in ibid., 318–48. These

sources and issues, as well as Singer's and Lifton's research on cults mentioned later, are discussed at greater length in Scott Selisker, "The Cult and the World System: The Topoi of David Mitchell's Global Novels," *Novel: A Forum on Fiction* 47, no. 3 (2014): 443–59.

9. Douglas E. Cowan and David G. Bromley, *Cults and New Religions: A Brief History* (Malden, Mass.: Blackwell, 2008), 216.

10. Lawrence S. Wright, *Going Clear: Scientology, Hollywood, and the Prison of Belief* (New York: Knopf, 2013).

11. See "Church of Scientology International Statement on Lawrence Wright's Book," January 11, 2013, http://www.scientologynews.org/statements. This statement links to the much more thoroughgoing website "How Lawrence Wright Got It So Wrong," http://lawrencewrightgoingclear.com/. Both allege that there are rampant factual inaccuracies in Wright, *Going Clear*.

12. Leigh Fondakowski, *Stories from Jonestown* (Minneapolis: University of Minnesota Press, 2013), 4–5, 25, 94–95.

13. Margaret Thaler Singer, *Cults in Our Midst: The Continuing Fight against Their Hidden Menace* (San Francisco: Jossey-Bass, 1995), 7. See also Selisker, "The Cult and the World System."

14. Robert Jay Lifton, *Thought Control and the Psychology of Totalism: A Study of "Brainwashing" in China* (1961; repr., Chapel Hill: University of North Carolina Press, 1989). For an early use of "totalism," see Erik H. Erikson, "Wholeness and Totality: A Psychiatric Contribution," in *Totalitarianism: Proceedings of a Conference Held at the American Academy of Arts and Sciences, March 1953*, ed. Carl J. Friedrich, 156–70 (Cambridge, Mass.: Harvard University Press, 1954).

15. Lifton, *Thought Control*, 420–30.

16. Lionel Trilling, *The Liberal Imagination* (New York: New York Review of Books Classics, 2008), xxi; qtd. in Lifton, *Thought Control*, 446. See also Selisker, "The Cult and the World System."

17. Eleanor Blau, "Ted Patrick Denies Assault in 'Kidnapping' Case," *New York Times*, March 31, 1973, 15.

18. Eleanor Blau, "Patrick's 'Deprograming' Is Recounted," *New York Times*, August 1, 1973, 14. At the time, "deprograming" and "programing," with one *m*, were the common spellings in the *New York Times*.

19. Andreas Killen, *1973 Nervous Breakdown: Watergate, Warhol, and the Birth of Post-Sixties America* (New York: Bloomsbury, 2007), 112.

20. Ted Patrick, with Tom Dulack, *Let Our Children Go!* (New York: Dutton, 1976), 67.

21. Elaine Scarry, *The Body in Pain: The Making and Unmaking of the World* (New York: Oxford University Press, 1987).

22. Ibid., 40. Scarry's overall argument likens the "unmaking" of the world in torture to the structure of war as unmaking the state.

23. For Scarry, *pain* is a word that is grammatically "objectless," like an intransitive verb; whereas "fear is fear of" something, pain is described in terms only of intensity and duration. Ibid., 34, 162, 5.

24. Ibid., 40–41, 44.

25. Patrick, *Let Our Children Go!*, 76.

26. Ibid., 76.

27. Ibid., 49.

28. "Kidnaping for Christ," *TIME*, March 12, 1973, 83–86.

29. Patrick, *Let Our Children Go!*, 77.

30. Ibid., 284.

31. Paul L. Montgomery, "19-Year-Old Member of Marxist Unit 'Deprogramed' and in Parents' Care: 'Zombie-Type Situation,'" *New York Times*, July 27, 1974, 59.

32. Joan Didion, "Girl of the Golden West," in *After Henry*, 95–109 (New York: Vintage, 1993); Nancy Isenberg, "Not 'Anyone's Daughter': Patty Hearst and the Postmodern Legal Subject," *American Quarterly* 52, no. 4 (2000): 639–81; William Graebner, *Patty's Got a Gun: Patricia Hearst in 1970s America* (Chicago: University of Chicago Press, 2008), 8. While all three of these accounts are convincing on their own terms, the later media fascination with John Walker Lindh in 2002 suggests that Hearst's trial carries interest beyond the periodization of the 1970s.

33. Carolyn Anspacher, ed., *The Trial of Patty Hearst* (San Francisco: Great Fidelity Press, 1976), 119.

34. Ibid., 314.

35. Ibid., 581.

36. Ibid., 77.

37. Such a strategy for describing violent acts bears comparison to Walter Benjamin's description of the "foundational" act of violence in his essay "Critique of Violence," in *Reflections: Essays, Aphorisms, Autobiographical Writings*, ed. Peter Demetz, 277–300 (New York: Schocken, 1969). The SLA claimed all their violent acts as foundational ones, in implicit comparison to the violent acts that had been necessary in the successful revolutions of the past.

38. Anspacher, *Trial*, 325.

39. Ibid., 314.

40. Patrick, *Let Our Children Go!*, 283.

41. Ibid., 284. Patrick ends his memoir in November 1975, with an account of his second offer of help to the Hearsts, after "batteries of psychiatrists and lawyers had intervened," but before Patty Hearst went to trial. Ibid. Indeed, the ongoing curiosity surrounding Hearst's captivity likely widened the market for Patrick's memoir.

42. Qtd. in Graebner, *Patty's Got a Gun*, 63.

43. Anspacher, *Trial*, 78.

44. Ibid., 119.

45. Qtd. in Graebner, *Patty's Got a Gun*, 91.

46. Anspacher, *Trial*, 581.

47. See, e.g., the trial's inclusion in Charles Patrick Ewing and Joseph T. McCann's *The Mind on Trial: Great Cases in Law and Psychology* (New York: Oxford University Press, 2006), 31–44.

48. Patricia Hearst, with Alvin Moscow, *Every Secret Thing* (New York: Doubleday, 1981), 206.

49. Ibid., 279.

50. Ibid., 69.

51. Ibid., 383.

52. Susan Choi, *American Woman* (New York: HarperCollins, 2003).

53. Pamela White Hadas, *Beside Herself: Pocahontas to Patty Hearst* (New York: Knopf, 1983).

54. Pamela White Hadas, "Patty Hearst: Versions of Her Story," in ibid., 199, 203–4, 212.

55. Graebner, *Patty's Got a Gun*. A book of similar scope uses *The Manchurian Candidate* as an academic introduction to the Cold War period: Matthew Frye Jacobsen and Gaspar Gonzalez, *What Have They Built You to Do? The Manchurian Candidate and Cold War America* (Minneapolis: University of Minnesota Press, 2006).

56. Christopher Sorrentino, *Trance: A Novel* (New York: Picador, 2006).

57. Ibid., 39.

58. Ibid., 181, 170.

59. Ibid., 86–87, emphasis original.

60. Ibid., 149.

61. Ibid., 149, 160.

62. The term *moral panic* was coined in Stuart Hall et al.'s landmark study of media and police culture, *Policing the Crisis: Mugging, the State, and Law and Order* (1978; repr., New York: Palgrave, 2013). In their study of anxiety about immigration, the authors studied the use of the term *mugging* in the media as a racially coded buzzword about crimes involving immigrants in the early 1970s in the United Kingdom. In a somewhat less pronounced way, the U.S. media's consistent use of the term *cults* capitalized on readers' anxieties about these groups and their dangers to middle-aged readers' children.

63. "Following the Leader: How Cults Lure the Drifting and Discontented—and Keep Them," *TIME*, December 11, 1978, 36.

64. "Cult Wars on Capitol Hill: Dire Warnings, and First Amendment Pleas," *TIME*, February 19, 1979, 54.

65. Ibid.

66. Amy Hungerford, *Postmodern Belief: American Literature and Religion since 1960* (Princeton, N.J.: Princeton University Press, 2010), 59.

67. Chuck Palahniuk, *Survivor* (New York: W. W. Norton, 1999); Palahniuk, *Fight Club* (New York: W. W. Norton, 1996).

68. Colin Hutchinson, in "Cult Fiction: 'Good' and 'Bad' Communities in the Contemporary American Novel," *Journal of American Studies* 42, no. 1 (2008): 35–50, argues the phenomenon of the cult in contemporary fiction as a means for progressives to reassess "communitarian themes and secessionist strategies" for a "dissenting liberal left" (36). In a counterintuitive reading of Katherine Dunn's novel *Geek Love* (1989), Hutchinson reads the inclusion of a mock cult ("Arturism," started by a huckster named Arty) as a satire on U.S. culture more broadly. "The implication here is that mainstream American society has become a form of 'super-cult': a 'bad' community in which liberty, responsibility and identity—even the ability to distinguish truth from fiction—are surrendered to a figurehead such as the President, or to an abstract entity such as the Constitution, and are returned as a set of reassuring myths and platitudes" (46). His argument resonates with that of *One Flew over the Cuckoo's Nest,* in which the closure of the mental ward is supposedly mirrored onto the larger "Combine" of society as a whole.

69. Hutchinson, ibid., reads *Mao II* in these terms, again with an eye to the possibility of collectivity, wherein groups do not share "truly collective sensibility" but instead cohere through the "surrender of collective and individual agency to the dominant will of a charismatic leader" (47). Hungerford, *Postmodern Belief,* reads *Mao II* alongside *The Names* with attention to language's fullness and emptiness, noting that "the cult . . . is never endorsed" in DeLillo's fiction (73).

70. Don DeLillo, *Mao II* (New York: Penguin, 1992), 6.

71. Ibid., 9.

72. Ibid., 79.

73. Ibid., 119.

74. Ibid., 179.

75. Ibid., 194.

76. Ibid., 224.

77. Ibid., 236.

78. Ibid., 16.

79. Isenberg, "Not 'Anyone's Daughter,'" 641.

80. Lance Morrow, "In the Name of God: When Faith Turns to Terror," *TIME,* March 15, 1993, 24–25.

81. Ibid., 24.

82. Ibid.

83. Ibid.

84. Samuel Huntington, *The Clash of Civilizations and the Remaking of the World Order* (New York: Simon and Schuster, 1996); Benjamin R. Barber, *Jihad vs. McWorld* (New York: Ballantine, 1995).

5. Fundamentalist Automatons

1. *Homeland,* created by Howard Gordon and Alex Gansa (New York: Showtime Networks Inc., 2011).

2. Chapter 4 traced how sociologists and psychologists of NRMs, including Robert Jay Lifton and Margaret Thaler Singer, borrowed from (sometimes their own) earlier descriptions of communist "brainwashing" and the features of the totalitarian state in describing the unfreedom that obtains within the cult as a result of a *form* of society. This *topos* of the cult emphasized specialized language, closed-off communications, and charismatic leadership as features that made a NRM a "cult" and, consequently, a site of false consciousness or pathology.

3. On *Homeland*'s Islamophobia, see Laila Al-Arian, "TV's Most Islamophobic Show," December 15, 2012, http://www.salon.com/2012/12/15/tvs_most_islamophobic_show/. Al-Arian is concerned with the details the show gets wrong (inaccurate depictions of Beirut, the character name "Raqim" coming from hip hop rather than Muslim culture); in my following discussion, I would argue that the show's Islamophobia would take shape mainly in its treatment of Islam as a pathology.

4. As of January 2014, *Falling Man* was a subject term in 59 articles, book chapters, or dissertation chapters in the *MLA International Bibliography*, out of the 291 such works devoted to DeLillo in that period since 2007. Compare to thirty-eight for another comparably popular American author at the time, David Foster Wallace, and to twenty in the 2007–13 period for his major work *Infinite Jest* (1996). Toni Morrison has 464 articles from the same period since *Falling Man*'s publication, and her contemporaneous work *A Mercy* (2008) has 51 hits. See later for a survey of the main currents in the scholarship on *Falling Man*.

5. John Updike, *Terrorist* (New York: Ballantine, 2007).

6. Bruce Holsinger, *Neomedievalism, Neoconservatism, and the War on Terror* (Chicago: Prickly Paradigm Press, 2007).

7. Raphael Patai, *The Arab Mind*, rev. ed. (New York: Scribner, 1983), 310, 209. Timothy Melley has traced how *Arab Mind* has "constitute[d] a mainstay text in diplomatic and military circles" after having been "reissued in November 2001" and then circulated widely between academics and neoconservative thinkers. Melley, *The Covert Sphere: Secrecy, Fiction, and the National Security State* (Ithaca, N.Y.: Cornell University Press, 2012), 83–84.

8. Leti Volpp, "The Citizen and the Terrorist," *UCLA Law Review* 49 (2002): 1576.

9. Jasbir Puar, *Terrorist Assemblages: Homonationalism in Queer Times* (Durham, N.C.: Duke University Press, 2007).

10. Leerom Medovoi, "Dogma-Line Racism: Islamophobia and the Second Axis of Race," *Social Text* 30, no. 2 (2012): 45.

11. Ibid.

12. Lindh was a white American, whereas Reid was a British citizen of Jamaican descent.

13. Alberto Toscano, *Fanaticism: On the Uses of an Idea* (New York: Verso, 2010). See also Timothy Marr, *The Cultural Roots of American Islamicism* (New

York: Cambridge University Press, 2006), which traces the widely varied and often admiring U.S. appropriations of Islamic culture during the nineteenth century. Between Marr's and Toscano's genealogies, there is a distinction to be made between Islam as a source of cultural and artistic phenomena, on one hand, and as a Western concept of an alien form of religiosity, on the other.

14. Toscano, *Fanaticism*, 110.

15. Ibid., 249.

16. Ibid., 154.

17. Qtd. in ibid., 221. In that essay, Arendt describes the "approach of the social sciences which," she claims, "treat ideology and religion as one and the same thing because they believe that Communism (or nationalism or imperialism, etc.) fulfills for its adherents the same 'function' that our religious denominations fulfill in a free society." See Hannah Arendt, "Religion and Politics," in *Essays in Understanding, 1930–1954: Formation, Exile, and Totalitarianism* (New York: Schocken, 2005), 372. For Arendt, this functionalist approach elides one important difference, namely, religion's sometime function of providing "freedom to be and remain outside the realm of secular society altogether" (373). That distinction fades in importance, however, in discussions where terrorist acts—meant to influence secular society—are in play. Again, Arendt's understanding of totalitarianism as an abstract "law"—of history for Marxists, of Darwinism for Nazis—brought into being in the world is structurally quite similar to Hegel's accounts of the French Terror and Islam. This abstractive law, in Arendt's terms in *Origins of Totalitarianism*, "claim[s] to transform the human species into an active carrier of a law to which human beings otherwise would only passively and reluctantly be subjected" (462).

18. Raphael Patai, *The Arab Mind*, rev. ed. (New York: Scribner, 1983), 310.

19. Representative texts include Deepa Kumar, *Islamophobia and the Politics of Empire* (Chicago: Haymarket Books, 2012); Carl W. Ernst, *Islamophobia in America: The Anatomy of Intolerance* (New York: Palgrave Macmillan, 2013); and, on the right, Walid Shoebat and Ben Barrack, *The Case FOR Islamophobia: Jihad by the Word; America's Final Warning* (New York: Top Executive Media, 2013), and David Horowitz and Robert Spencer, *Islamophobia: Thought Crime of the Totalitarian Future* (New York: David Horowitz Freedom Center, 2011).

20. John L. Esposito and Ibrahim Kalin, *Islamophobia: The Challenge of Pluralism in the 21st Century* (New York: Oxford University Press, 2011), xxiii.

21. Lisa Stampnitzky, *Disciplining Terror: How Experts Invented "Terrorism"* (New York: Cambridge University Press, 2013), 3.

22. Jacqueline Rose, "Deadly Embrace," *London Review of Books* 26, no. 21 (2004): 24.

23. Puar, *Terrorist Assemblages*, 52. See also Judith Butler, "Explanation and Exoneration, or What We Can Hear," *Social Text* 72 (Fall 2002): 177–88.

24. Mark Claywell, dir., *American Jihadist* (Philadelphia: Breaking Glass Pictures, 2011).

25. In the neo-Freudian register, one of Ali's spiritual guides makes the commonsensical explanation, "When you grow up getting beat up, you got a complex, okay; that complex manifests itself in a real reaction." The journalist Tod Robberson appears as a talking head to speculate, "The real Isa is the one who was getting beaten up for his lunch money." Ali himself recounts having a suicidal epiphany as a young war veteran, which seems like an equally convincing psychological account of his propensity for violence. The documentary does not adjudicate between this perspective and the CIA expert's account of Isa's consciousness.

26. Mohsin Hamid, *The Reluctant Fundamentalist* (New York: Mariner Books, 2007). A film adaptation was made in 2012, directed by Mira Nair.

27. Leerom Medovoi, "'Terminal Crisis?' From the Worlding of American Literature to World-System Literature," *American Literary History* 23, no. 3 (2011): 645.

28. Hamid, *Reluctant Fundamentalist*, 166. In "Terminal Crisis," Medovoi characterizes Erica as a "heavy-handed allegorical figure" who "personifies the narrator's romantic national investments, his ill-fated love affair with American life" (654).

29. *Battlestar Galactica*, created by Ronald D. Moore, British Sky Broadcasting, and NBC Universal Television (Universal City, Calif.: Universal Studios Home Entertainment, 2005–9).

30. Ronald D. Moore has since said that the initial decision to create only humanoid Cylons was a budgetary one, because the metallic Cylons would require expensive photorealistic computer-generated imagery. Annalee Newitz, "*Battlestar Galactica*'s Humanoid Cylons Were a Result of Budget Limits," July 20, 2013, http://io9.com/battlestar-galacticas-humanoid-cylons-were-a-result-of -844621846.

31. One episode in particular has garnered critical attention for its staging of the Abu Ghraib scandal. In this episode, "Pegasus," in the second season, the protagonists encounter another ship that survived the Cylon attack, but it quickly becomes clear that they have survived not with compassion but through cruel and authoritarian measures that have included the brutal torture and rape of a Cylon prisoner. See Karen Randell, "'Now the Gloves Come Off': The Problematic of 'Enhanced Interrogation Techniques' in *Battlestar Galactica*," *Cinema Journal* 51, no. 1 (2011): 168–73.

32. *Battlestar Galactica*'s creators also produced a spin-off series, *Caprica*, which addressed similar issues with a stronger emphasis on information technology. In that plot, a young girl named Zoe is able to create a human construct of herself using her data from social media—as if to suggest that in the age of social media, we and our children could be reduced to those sets of data, a version of what David Golumbia has discussed as the "upload fantasy." That data plus a mysterious gurgle in the machine create a computer-embedded construct. The associative tableau of *Caprica* includes virtual reality games,

advanced data-mining AI, an embattled ethnic minority with mafia ties, and a monotheistic terrorist cell as a suggestive expansion of the *Battlestar Galactica* franchise, but *Caprica*'s comparatively short two-season run included few other conceptual developments beyond the original show. See David Golumbia, "The Future of New Media: Embodying Kurzweil's Singularity in *Dollhouse, Battlestar Galactica*, and *Gamer*," in *Media Studies Futures*, vol. 6 of *The International Encyclopedia of Media Studies*, ed. Kelly Gates and Angharad N. Valdivia, 479– 502 (Malden, Mass.: Blackwell, 2013).

33. A notable exception would be Peter Watts's *Blindsight* (New York: Tor, 2006), in which the alien is the most radically nonhuman consciousness possible, a scenario not unlike that of Stanisław Lem's *Solaris* (1961; repr., New York: Mariner, 2002). Otherwise, the "Borg" of *Star Trek: The Next Generation* seem to be the last of the implacably inhuman aliens.

34. Michiko Kakutani, "John Updike's 'Terrorist' Imagines a Homegrown Threat to Homeland Security," *New York Times*, June 6, 2006, E1.

35. The concerns of mourning, the representability of the event, and trauma theory more broadly are dominant in much of the early scholarship on *Falling Man*, such as the collected volume of Ann Keniston and Jeanne Follansbee Quinn, eds., *Literature after 9/11* (New York: Routledge, 2008), and elsewhere. With these, see, e.g., Linda S. Kauffman, "The Wake of Terror: Don DeLillo's 'In the Ruins of the Future,' 'Baader-Meinhof,' and *Falling Man*," *MFS: Modern Fiction Studies* 54, no. 2 (2008): 353–77. Kaufmann characterizes the main concern of the novel as "the repression of memory and the memory of repression" (353), though her essay also works to situate *Falling Man* in continuity with *Mao II* and with DeLillo's story "Baader-Meinhof" and his 2001 essay "In the Ruins of the Future." Kristiaan Versluys, in *Out of the Blue: September 11 and the Novel* (New York: Columbia University Press, 2009), contextualizes the story among contemporary arguments about 9/11 as he reads it through the Freudian lens of melancholia. Rachel Greenwald Smith, in "Organic Shrapnel: Affect and Aesthetics in 9/11 Fiction," *American Literature* 83, no. 1 (2011): 153–74, discusses *Falling Man* in the course of an application of affect theory to the fiction of 9/11; Laura E. Tanner, in "Holding onto 9/11: The Shifting Grounds of Materiality," *PMLA* 127, no. 1 (2012): 58–76, reads the text with regard to its interest in mediation; Ahmed Gamal, in "Encounters with Strangeness in the Post-9/11 Novel," *Interdisciplinary Literary Studies* 14, no. 1 (2012): 95–116, discusses *Falling Man* and Updike's *Terrorist* relative to tropes of Orientalism. See also Aaron DeRosa, "Alterity and the Radical Other in Post-9/11 Fiction: DeLillo's *Falling Man* and Walter's *The Zero*," *Arizona Quarterly: A Journal of American Literature, Culture, and Theory* 69, no. 3 (2013): 157–83; Katrina Harack, "Embedded and Embodied Memories: Body, Space, and Time in Don DeLillo's *White Noise* and *Falling Man*," *Contemporary Literature* 54, no. 2 (2013): 303–36; Elizabeth S. Anker, "Allegories of Falling and the 9/11 Novel," *American Literary History* 23, no. 3 (2011): 463–82; and Hamilton

Carroll, "'Like Nothing in This Life': September 11 and the Limits of Representation in Don DeLillo's *Falling Man*," *Studies in American Fiction* 40, no. 1 (2013): 106–30. This list touches on all the major journals devoted to contemporary U.S. literature, and as a rule, each of these essays on *Falling Man* discusses the novel's relationship to ethics, memorialization, literary representation, or the representation of the terrorist, often pairing *Falling Man* with one or more less well-known literary texts about 9/11.

36. DeLillo, *Falling Man*, 231.

37. Ibid.

38. Ibid.

39. Ibid.

40. Updike, *Terrorist*, 3.

41. Ibid., 6.

42. In a more positive review in *New York Magazine*, June 5, 2006, John Leonard wrote that "the characters in *Terrorist* may be sketchy, and the action perfunctory, and the stereotyping wearisome, but Ahmad stirs up sediment in us" (71). The literary critic Anna Hartnell, while she finds *Terrorist*'s "exploration of the relationship between faith and violence . . . problematic," praises the novel for its refusal of "national triumphalism" in the wake of 9/11. Hartnell, "Violence and the Faithful in Post-9/11 America: Updike's *Terrorist*, Islam, and the Specter of Exceptionalism," *MFS: Modern Fiction Studies* 57, no. 3 (2011): 489.

43. In Updike's allegory, we see another project for the novel of terror, embodied in the guidance counselor protagonist Jack. The high school guidance counselor acts as a sort of moral compass for the student, as the single school official who concerns himself entirely, and holistically, with students' futures. As an older and near-obsolete white male protagonist, this guidance counselor might be said to resemble Updike himself, in a dim view of the novelist's prospects for serving as a moral compass for others.

44. DeLillo, *Falling Man*, 68.

45. Toscano, *Fanaticism*, 154.

46. Lianne's character also pursues the religious philosophy of Søren Kierkegaard, and Daniel Greenspan and James Gourley have both pursued the way in which Kierkegaard's figure of the knight of faith (not directly mentioned) provides a way of imagining Hammad's religious consciousness. Greenspan, "Don DeLillo: Kierkegaard and the Grave in the Air," in *Kierkegaard Research: Sources, Reception, and Resources: Vol. 12, Tome IV*, 81–100 (New York: Ashgate, 2013); Gourley, *Terrorism and Temporality in the Works of Thomas Pynchon and Don DeLillo* (New York: Bloomsbury, 2013), 71–72.

47. *Falling Man*'s subplot in which Alzheimer's patients tell stories about the event could be considered as a concession to this convention, but Keith's course through the novel is much less therapeutic.

48. DeLillo, *Falling Man*, 91.

49. Ibid., 99.

50. Ibid., 211.

51. Richard Grusin, in *Premediation: Affect and Mediality after 9/11* (New York: Palgrave Macmillan, 2010), compares the poker to ongoing, PTSD-esque repetition, in which his "enslavement to the Las Vegas poker tournament-machine is emblematic for DeLillo of life in a post-9/11 United States" (107).

52. Ibid., 226.

53. Ibid., 230.

54. Ibid., 226.

55. Ibid., 174.

56. Along the lines of "cult" consciousness, Hungerford, *Postmodern Belief*, has argued that the religiosity of DeLillo's text is an embrace of transcendence through language. She claims that "the cult . . . is never endorsed" in DeLillo's earlier fiction, and his depictions of the Names cult in *The Names* and of the Moonies in *Mao II* would seem to be of a piece with other popular representations of cults over the last twenty-five years (73). Hungerford does not address *Falling Man*, but she claims that in "In the Ruins of the Future," DeLillo "indicts fundamentalism for its failure to see the human person as such and to value her" (75), an assessment with which I agree.

57. Melley, *Covert Sphere*, 201–2.

58. In a forum on the visual culture of the War on Terror, Matthew Delmont, in "Introduction: Visual Culture and the War on Terror," *American Quarterly* 65, no. 1 (2013): 157–60, raises one version of this quandary relative to the show, noting that "Obama's praise for Homeland is intriguing because the show offers a critical view of drone strikes, which have become an increasingly important and controversial weapon in the war on terror under the Obama administration" (158).

59. While the film updates the original's technologies to include an implanted computer chip in Captain Marco's shoulder and modifies the film's ending so that Marco's resistance is revealed as part of the plan, and he takes the original Shaw's place as the assassin in the final scene. Its incorrect usage of facially applied henna paintings on veiled dark women is as close as the film comes to representing Islam, Shaw's and Marco's automatism playing instead as being corporate pawns. For an international survey of henna painting traditions, see Loretta Roome, *Mehndi, the Timeless Art of Henna Painting* (New York: St. Martin's Press, 1998).

60. *The Conversation*, dir. Francis Ford Coppola, American Zoetrope (1974; Burbank, Calif.: Buena Vista Home Entertainment, 2005).

61. The first of these scenarios exemplifies Jasbir Puar's arguments about the constitutive queerness of the Muslim and a self-contradictory structure of "U.S. sexual exceptionalism": here a logic of homophobic prejudice condemns the queer Muslim male even as the CIA delights in calling all Muslims homophobic. See Puar, *Terrorist Assemblages*, 3. The American CIA interrogators in

Homeland even delight in the supposed misogyny of the Saudi Arabian diplomat, whom they assume "will be thrown if [the interrogator in the interview] is a woman." With these bathhouse photos of Al-Zahrani's illicit gay affair, Carrie and company threaten to ruin not only his family life but his public reputation "back home." A complex logic of name-calling allows the CIA to imply that Muslims at large are homophobic while taking unrepentant glee in exposing the upstanding Muslim man as a homosexual—in one odd line of dialogue, he's 'a "naughty boy"—and in showing him for what he really is. Moreover, that constitutive queerness of the Muslim dovetails here with a jab at the repressive form of Muslim society in its multifaceted unfreedom of thought and self-expression. The text places this strain of homophobic Islamophobia on display with plausible critical distance, considering that the failure of the related operation will highlight a failure of both means and ends for the investigative team.

62. Volpp, "Citizen as Terrorist," 1576.

63. Jeremy Egner, "A World War II Soldier Enters the Post-Iraq Age," *New York Times*, December 9, 2011, AR6. In the interview, Lewis also states, "I had one stipulation when I accepted the role, which was that there would be no lazy, easy parallels drawn between violence and Islam. Because I felt that would be irresponsible." This statement also articulates the demand from Kakutani's review of *Terrorist* that contemporary representations of terrorists be made through round, well-developed characters. Lewis also provides here one criterion through which we might judge whether the show is Islamophobic.

64. The four distinct suspects widely associated with brainwashing in the post-9/11 moment are discussed in Don Oldenburg, "Stressed to Kill: The Defense of Brainwashing; Sniper Suspect's Claim Triggers More Debate," *Washington Post*, November 21, 2003, C1. See the introduction for a longer discussion of Walker Lindh's portrayal in the U.S. news media.

65. Remote programming is a leitmotif of *Homeland*'s second season as well. This scene of indoctrination is paralleled in a widely panned subplot late in the second season, in which Abu Nazir assassinates the vice president by hacking wirelessly into his pacemaker, with Brody's assistance.

66. John Lahr, "Varieties of Disturbance; Where Do Claire Danes's Volcanic Performances Come From?," *New Yorker*, September 9, 2013, 50. The profile highlights the bodily dimensions of Danes's most intense acting, which sees the actress work herself into fits of shaking and weeping.

67. In a long arc early in *Homeland*'s third season, the show presents comparative institutionalizations of Carrie (in a mental ward), Dana (Brody's daughter, who attempted suicide), and Brody (in hiding, under the care of a suspicious doctor who keeps him in a heroin-induced haze). Much like the anti-institutional fictions discussed in chapter 3, these institutional dynamics foreground the defiance and freedom of Dana and Carrie in particular, while Brody succumbs to his drug-induced stupor. I read this plot arc as an attempt,

after the death of the primary villain Nazir at the end of the second season, to put the characters' extreme psychological states back on display, as they had been in the first season.

Conclusion

1. See Michael Rogin, *Ronald Reagan, the Movie: And Other Episodes in Political Demonology* (Berkeley: University of California Press, 1988).

2. Kay, *Who Wrote the Book of Life?*, xviii.

3. Latour, *Reassembling the Social*, 10.

4. Ibid., 37.

5. Suarez, *Daemon*, 613.

6. Or, as Rita Felski explains it, "the 'actor' in actor-network theory is not a self-authorizing subject, an independent agent who summons up actions and orchestrates events. Rather, actors only become actors via their relations with other phenomena, as mediators and translators linked in extended constellations of cause and effect." Felski, "Context Stinks!," *New Literary History* 42, no. 4 (2011): 583.

7. Bruno Latour, "Why Has Critique Run Out of Steam? From Matters of Fact to Matters of Concern," *Critical Inquiry* 30, no. 2 (2004): 225–48.

Index

Friedan, Betty, 6, 33, 74, 99, 107, 108, 182, 189, 206n4, 216n92, 216n99; consumerist patriarchy and, 92; dehumanization and, 93; feminist rhetoric of, 91; household control and, 94; housewife false consciousness and, 93; POW imagery of, 94; "Sexual Sell" and, 92; total institution and, 95
Friedkin, William, 131
Fromm, Erich, 15, 39, 46, 49, 53, 57, 70, 200n10, 206n11, 216n102; anti-totalitarianism and, 91; fascism and, 94; human automaton and, 4; totalitarianism and, 40
Fukuyama, Francis: end of history of, 150
Fuller, Samuel, 33, 52, 202n61
Fu Manchu, 113. *See also Mask of Fu Manchu, The* (film)
fundamentalism, 31, 101, 124, 159, 177, 182, 223n82, 233n56; automatism and, 124; Islamic, 153; religious, 7; totalitarian, 161
Fung, Catherine, 114

Gandhi, Vikram, 126
Garland, Alex, 14, 111
Garveyism, 77
Gattaca (film), 105
Geek Love (Dunn), 227n68
genetics, 34, 100, 105, 219n20
Gilbreth, Frank Bunker, Sr., 102
Ginsberg, Allen, 72, 76, 206n8
Goffman, Erving, 206n6, 207n11, 220n42; total institution and, 73, 206n11
Going Clear (Wright), 129
Golumbia, David, 105, 230n32
Gourley, James, 232n46
governance: totalitarian, 46, 47
Graebner, William, 135, 144
Gramsci, Antonio, 134

Great Dictator, The (film), 26
Greenberg, Clement, 13
Greening of America, The (Reich), 90–91
Greenspan, Daniel, 232n46
Greif, Mark, 8–9, 113
Grey, Joe: photo of, 62
Grundwerk (Kant), 196n59
Guantánamo, 168
Guattari, Félix, 15
Guerilla: The Taking of Patty Hearst (documentary), 144
Gulf War syndrome, 204–5n89
Gunning, Tom, 195n51
Gutman, Jeremiah, 147

Hadas, Pamela White, 144
Haiti, U.S. occupation of, 22, 197n66
Hall, Stuart: moral panic and, 226n62
Halliburton, 168
Halperin, Victor Hugo, 14
Hamid, Mohsin, 158, 159
Hamlet (Shakespeare), 57, 66
Haraway, Donna, 8, 107, 115–16, 117, 118; cyborg of, 116, 124
Hardt, Michael, 150
Hare Krishnas, 128, 139, 144, 146, 147
Harlem, 77, 82, 84, 89
Harris, Joel Chandler, 87
Hartnell, Anna: on *Terrorist*, 232n42
Hartsock, Nancy, 107
Harvey, Lawrence, 172; photo of, 62, 65
Hassan, Steven: on brainwashing, 2
Hayek, Friedrich, 30
Hayles, N. Katherine, 109, 116–17, 223n82
Hearst, Patty, 4, 127, 143, 145, 170, 189n2; assessment of, 150; brainwashing and, 2, 135–37, 139–42;

Index

one-dimensional, 162; terrorist, 161

Rockefeller, John D., 89

Rockwell, Norman, 30

Rogin, Michael: demonization and, 180

Roosevelt, Franklin D., 30

Rorty, Richard, 123, 222n81

Rose, Jacqueline, 157, 177

Rosemary's Baby (film), 144, 216n100

Rosenberg, Melissa, 14

Rosenblueth, Arturo, 103

Rosicrucians, 128

Ross, Andrew, 81, 214n79

Ross, Katherine, 94

Ross, Marlon, 209n25

R.U.R. (Čapek), 7

Ryle, Gilbert, 169

Said, Edward, 154

Saint-Léon, Arthur, 12

Sambo doll, scene, 85, 86–87, 95

samurai films, 221n54

"Sandman, The" (Hoffmann), 12, 18, 19, 196n58

Sapir–Whorf hypothesis, 200n22

Sartre, Jean-Paul, 190n17

Saturday Evening Post, The, 30, 54

Scarry, Elaine, 131–32

Schlemmer, Oscar, 12

Schrader, Paul, 144

science, 79, 127, 141, 152; civilization and, 49; technology and, 5

science fiction, 5, 34, 35, 92, 98, 100, 102, 111, 112–13, 117, 123, 162, 217n1; contemporary, 161; feminist, 116; films, 14; propaganda and, 4; technology and, 116

scientific knowledge, 32, 37, 152, 182

scientific management, 11, 81

scientific paradigms, 32, 180

Scientology, 129, 146

Scott, Jack, 146, 149

Scott, Ridley, 96, 110, 114, 159

Scott, W. Richard, 206n6

"SCUM Manifesto" (Solanas): automation and, 107

Seabrook, William, 22, 23, 38–39

Seed, David, 15, 47, 60, 106, 219n25

self: dialectic of, 179; human, 161; transhuman, 117

self-definition: democratic, 130

self-determination, 73, 142

Selfish Gene, The (Dawkins), 121

self-possession: liberal humanism and, 190n17

Sellers, Peter, 29; photo of, 29

Seltzer, Mark, 13, 80

Seventh Day Adventists, 126

sexual abuse, 126, 135

Shadow and Act (Ellison), 78

Shannon, Claude, 103, 105, 120

Shelley, Mary, 13

Shih, Shu-mei, 9, 191n19

Ship Who Sang, The (McCaffrey), 116

Shklovsky, Victor, 42, 201n25

Shock Corridor (film), 33, 52

Siegel, Don, 95

Silicon Valley, 34, 96, 98, 217n107

Silko, Leslie Marmon, 223n81

Simondon, Gilbert, 220n42

Simulacra, The (Dick), 113

Sinatra, Frank, 5, 58, 204n86; photo of, 62

Singer, Margaret Thaler, 129, 136, 141, 142, 224n8, 228n2; brainwashing and, 140; expert testimony of, 137, 139; traumatic neurosis and, 137

Singh, Nikhil Pal, 21, 51, 209n23, 210n28

Skinner, B. F., 38, 104, 201n32

SLA. *See* Symbionese Liberation Army (SLA)

slavery, 5, 196n58

Slave World, 26, 27 (fig.), 29, 31, 70

SCOTT SELISKER is assistant professor of English at the University of Arizona.